CHIME CLOCK REPAIR

STEVEN G. CONOVER

Steven G. Conover

CLOCKMAKERS NEWSLETTER, READING

Dedicated to my wife, Karen, for her encouragement, advice, and understanding.

Published by Clockmakers Newsletter
203 John Glenn Avenue
Reading, PA 19607

Second Edition, 1997
Text and Drawings Copyright © 1997 by Steven G. Conover

No part of this book may be reproduced in any manner without written permission from the publisher.

ISBN 0-9624766-6-8

CHIME CLOCK REPAIR

SECOND EDITION

STEVEN G. CONOVER
Author of CLOCK REPAIR BASICS and STRIKING CLOCK REPAIR GUIDE

Other Books by Steven G. Conover:
Clock Repair Basics $22.95
Striking Clock Repair Guide $22.95
Building an American Clock Movement $21.95
How to Repair Herschede Tubular Bell Clocks $12.95
Book prices include postage when ordered through:
    Clockmakers Newsletter
    Steven G. Conover, editor
    203 John Glenn Ave.
    Reading, PA 19607
    (610) 796-0969

Newsletter
Clockmakers Newsletter (monthly) $40.00/year or $70.00/2 years
Sample issue $3.00.

Video by Steven G. Conover: *Repairing Eight Day American Striking Clocks*, a Zantech Video Production, presents a fascinating look at count wheel movements, featuring examples by Gilbert, Sessions, and others. The video is offered through Clockmakers Newsletter at $39.95 plus $3.50 shipping.

CONTENTS

Introduction iv
1 Chime Basics 1
2 Common Repair Problems 7
3 Seth Thomas No.124 16
4 Seth Thomas No.113 26
5 Seth Thomas Sonora Chime 30
6 New Haven.. 36
7 Sessions Two-Train Chime 44
8 Waterbury Double Deck Chime................................. 52
9 Ansonia... 59
10 Globe 66
11 Winterhalder 69
12 Jacques... 75
13 Junghans... 81
14 Hermle 85
15 Urgos 06-Series... 93
16 Urgos 9-Tubular Bell... 97
17 Kieninger KSU/RSU 104
18 Gebr. Jauch.. 112
19 Mauthe W500... 116
20 Smith's... 120
   Index.. 125
   Acknowledgements... 126

INTRODUCTION

A chime clock is the most prized of our household timekeepers, for it does far more than just indicate the time on a dial. A chime clock sounds cathedral melodies, such as Westminster or Whittington, on tuned rods or tubular bells each quarter hour. Then it tolls the hour, sometimes with several notes struck in harmony.

Late in the 19th century, ornate grandfather clocks—some too tall for today's eight-foot ceilings—graced the homes of the wealthy. Leather tipped hammers played chime melodies on sets of long, nickel-plated tubular bells hanging in the cases. During the first years of the 20th century, clock manufacturers began mass-producing mantel and wall models with rod chimes, priced within the means of more people. In addition to German makers such as Junghans and Gustav Becker, there were a number of American clock companies such as Seth Thomas, Ansonia, Waterbury, and New Haven, each manufacturing a broad line of chime clocks. Modern-day clock companies—Howard Miller, Sligh, Ridgeway, to name a few—are really case designers who import German movements for their clocks. The remaining American chime clock manufacturer, Herschede, shipped its last tubular bell grandfather clock in 1984.

From the first, chime clocks have posed challenges for repairers. Complicated-looking movements have discouraged many amateurs and hobbyists, even those with a strong desire to learn. Professional clockmakers also have problems with chime clocks: they cannot afford to spend an inordinate amount of time studying, assembling, and adjusting them. With so many different chime movement designs in existence, even a busy professional may see some models only a few times each year, or even less often. Most have experienced the frustration of knowing they are "re-inventing the wheel" each time they take apart and restore a type of movement they have not worked on for a long time.

The unfortunate result is that a chime clock often receives only a small amount of tinkering and adjustment. Usually this does not work, even when the movement is completely cleaned as part of the effort. Modern ultrasonic equipment will clean a fully assembled movement very well, providing a temptation for those who would like to cut corners. What is really required is disassembly and repair done the "old-fashioned way" to restore the chime clock to good-running condition again.

More than ever, the public values mechanical clocks and is willing to pay for good, thorough repair work. Generations-old clocks are uncovered in the attic; modern clocks need periodic servicing as well. And with the increasing numbers of new chime clocks in use, skilled repairers will be even busier.

Chime Clock Repair is a reference work for those who restore chime clocks. I hope it will encourage the amateur to become more involved in the craft, building skill and confidence. For the professional clockmaker, these chapters will save valuable time by refreshing his or her knowledge of the movements. It is my hope that the book will spend more time on the repair bench, helping the clockmaker along, than it does in the bookcase.

We begin with two chapters on the basics of chime repair and common problems encountered at the bench. Each of the other chapters features a particular movement. I have long felt that the repairer needs detailed information on specific clocks, such as the Seth Thomas No. 124, in addition to general material concerning clock types, i.e. German chime, French strike, and so on.

With these thoughts in mind I have chosen a group of chime movements which are, with the possible exception of the Globe chime, not rare. The group includes movements from the early years of the 20th century as well as models still in production. The clocks represent a cross-section of the variety of chime work the repairer can expect to see at the bench. The Herschede American-made tubular bell movement does not appear in this work because I devoted a separate book to it: *How to Repair Herschede Tubular Bell Clocks*, also published by Clockmakers Newsletter.

The original *Chime Clock Repair* (1990) was the end result of countless hours spent studying, repairing, drawing, and writing about the movements over a period of ten years. I extensively edited, reorganized, and expanded the material from time to time as more information was obtained, adding new and unpublished illustrations and text wherever they would be helpful.

In contrast to the 1990 hardcover, this revised second edition is bound in soft cover, workshop manual style. The larger page format has placed the illustrations closer to the matching text. A few corrections were made along the way. The chapters were edited, and Chapter 3, in particular, was reorganized. In addition, I added photos and mainspring sizes for several of the movements.

Steven G. Conover
Reading, Pennsylvania, October, 1997

1

CHIME BASICS

WHY DISASSEMBLE?

Chime clocks are considered by many to be too complicated for repair. This complaint comes often from amateurs and collector-repairers who are interested in having fun with clocks. But it is also heard from professional clockmakers who find they are spending too much repair time on chime clocks to make them profitable. In an attempt to avoid problems and save time, many clockmakers oil and adjust the movements without actually taking them apart.

A basic premise of this book is that chime clocks should almost always be taken apart for repair. Disassembly is never a waste of time, and many "problem" chime clocks are only troublesome because they have not been taken apart and repaired properly. Frequently the clockmaker can spot trouble from outside a movement, which gives him a reason to take the movement apart. But just as often, there are hidden defects.

Pivot surfaces, in particular, are hidden. The clockmaker is only alerted to pivot problems if the hole is also worn. But all too often, the hole retains a normal appearance despite the fact that the pivot looks like the one in Figure 1. Over the years, dirt and oil in the pivot hole combine to form an abrasive. As the pivot turns in the hole, it becomes worn with concentric grooves. Sometimes the hole is not worn seriously enough to attract the repairer's attention as he inspects the movement. Yet the grooved

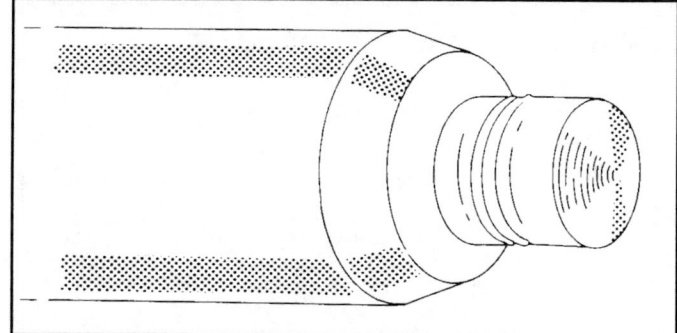

Fig. 1. Pivot worn with grooves.

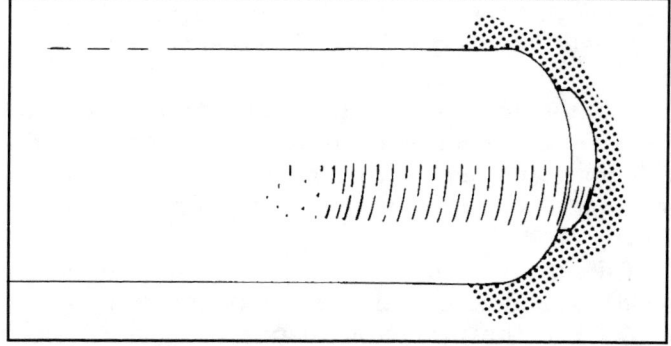

Fig. 2. Deposits accumulate between the pivot shoulder and the inside of the clock plate.

pivot resists front-to-back movement between the plates, increasing friction and almost eliminating endshake. If he has ultrasonically cleaned the assembled movement, the clockmaker can still detect the bad pivot by checking the endshake on the arbor, with the power off. But having already cleaned and dried the assembled movement, he is not looking for reasons to take the clock apart, and so will miss it.

Another hidden problem is created by improper ultrasonic cleaning of assembled movements. The process often leaves spots of greenish corrosion or hardened black oil and dirt in hard-to-reach places like the space between the pivot shoulder and plate (Figure 2). This happens because the clockmaker is trying to preserve the lacquer. He turns off the ultrasonic cleaner after a "safe" cleaning time, as soon as the assembled movement starts to look clean. He knows that if the lacquer does begin to lift, then it must be completely cleaned off the plates and wheels. Thus the cleaning is placed secondary to the outward appearance of the movement.

These "hidden" dirt and oil deposits stop clocks. Obviously, a cosmetic cleaning job accomplishes nothing. It isn't the dirt on the outside of the plates that causes clocks to fail: it is contamination on wheels and pinions and in the pivot holes. The substance may act almost like glue, sticking the pivot

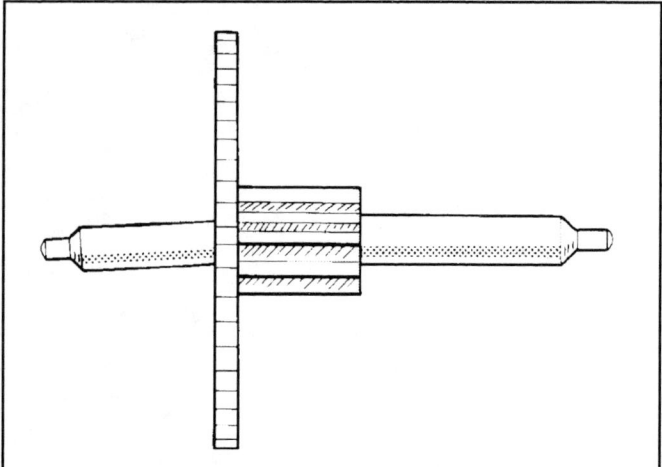

Fig. 3. A bent arbor is easy to miss, unless the movement is disassembled and each arbor checked.

Fig. 4. Typical pin barrel.

shoulder to the inside of the plate. Friction between the pivot and bearing surface will be high. New oil will not keep the pivot turning smoothly in the hole for long. It is critical to get at this deposit and remove it.

Take the movement apart to verify that cleaning, whether manual or ultrasonic, has removed all deposits from the movement. Inspect pivots and holes at the same time. Remember, ultrasonic cleaning is not a substitute for proper repair. It is simply a very effective way to clean clocks. No matter how clean the movement as a whole may look, it is only as clean as the worst spot of hidden dirt.

Consider also the bent pivot or arbor which is so hard to find in an assembled movement. It may be easy enough to see a chime or strike fly wobbling about, a sure sign of damage. But second and third arbors are harder to check. They turn very slowly, making it impossible to see the wobbling motion. And there always seems to be a mainspring barrel or a hammer assembly blocking your line of sight. Figure 3 represents a second wheel from a modern movement, bent from the full force of a broken mainspring, or a click released by someone who didn't use a let down key. The arbor is soft enough to absorb the shock by bending. So instead of a ruined pinion or a snapped pivot, there is a bent arbor. The clock will run for a day or two under full power. Then it stops as the off-center pinion jams against the barrel. Damage like this is easy to see in individual arbors after disassembly, but surprisingly hard to spot from outside, by looking at the complete movement.

CHIME MELODIES

Many repairers are afraid they won't be able to get the melodies synchronized again after repairing the movement. It is important to have an understanding of chimes so this will no longer be an obstacle. First of all it is important to realize that the chime melodies themselves do not require any adjustment. The location and sequence of the pins driven into the pin barrel will determine the melody to be sounded. Figure 4 is a typical pin barrel design, one of many variations.

Westminster chimes remain the standard for chime clocks whether early tubular bell, 1920's tambour or modern grandfather. Some clocks are dual or triple chime with the additional chime melodies to be selected by means of a lever on the dial. But at least Westminster is usually there. Knowing this can help you to synchronize the chimes when you are unfamiliar with the other chimes the clock plays.

After assembling the movement, make sure the melody starts at the right place. The pin barrel is like a phonograph record played once each hour: the record must start at the right place every time. For example, the Westminster chime plays four descending notes at the first quarter, an easy pattern to recognize. If the clock is put together without regard for this, the chime melody will begin at another point. You may hear four notes at the first quarter, but they will be the notes from another part of the tune. Most movements have an automatic

Chime	Notes per Sequence	Number of Notes at Hour	Number of Notes per Hour
Westminster	4	16	40
Winchester	6	24	60
Canterbury*	6	24	60
Whittington	8	32	80
St. Michael	8	32	80

* Herschede only

Fig. 5. Table of chime notes

CHAPTER 1 - CHIME BASICS

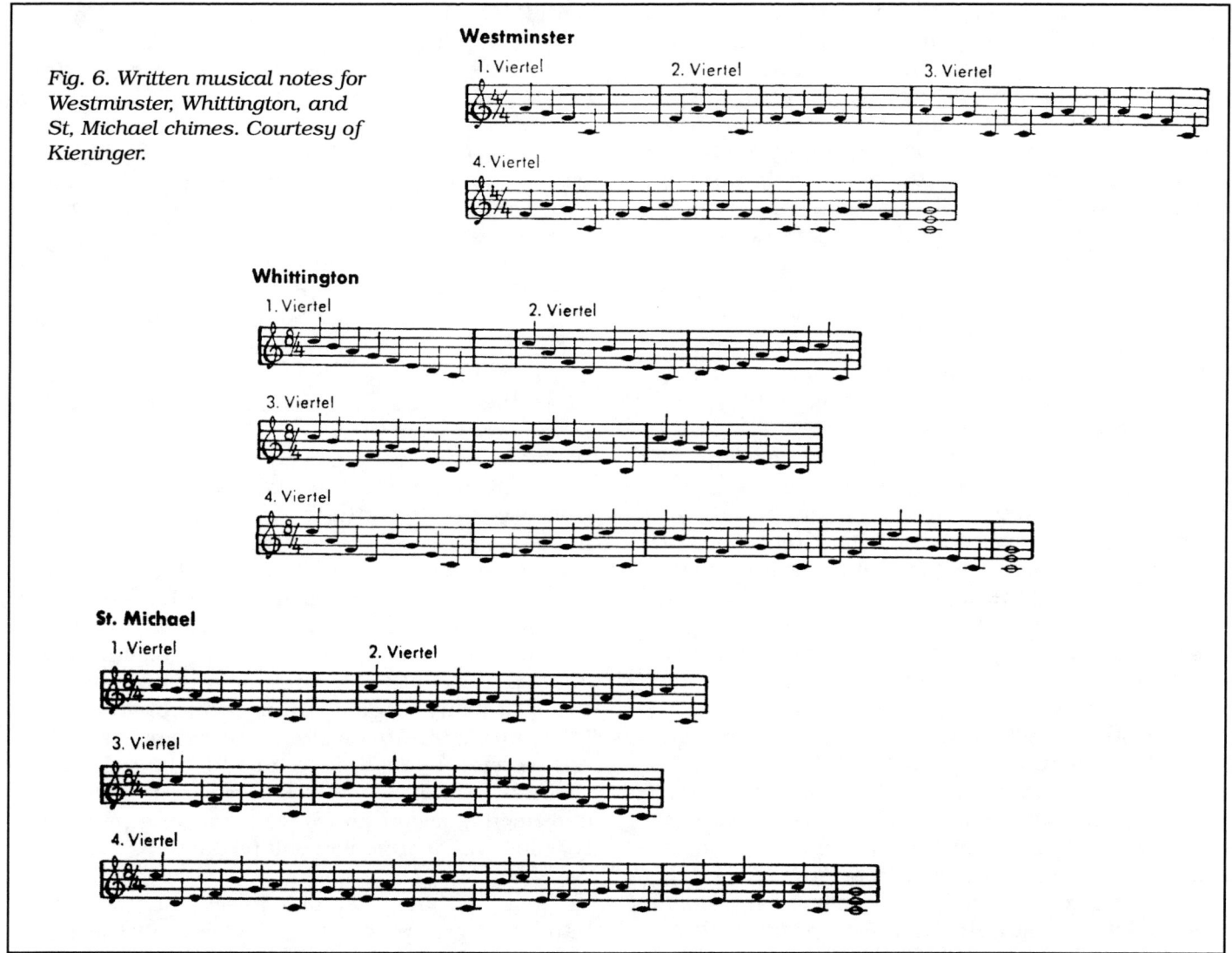

Fig. 6. Written musical notes for Westminster, Whittington, and St. Michael chimes. Courtesy of Kieninger.

chime correction device. But with few exceptions, all the mechanism does is to bring the chime melody back into synchronization with the minute hand. If the clock has been assembled the wrong way, the automatic chime correction feature will just keep returning the clock to the same unsynchronized chime melody.

How is it possible to recognize the various chime melodies in clocks? Figure 5 is a table of some of the common chimes in modern clocks. Just from the number of notes played, it is easy to distinguish one chime from another in some clocks. The dial is usually marked with the chime names, so it is not difficult to locate Westminster or some other chime if the mechanism is out of adjustment. Most clocks have a pause or rest between each measure, so you can count groups of notes easily. Or you can count the notes played in one hour, which is usually one revolution of the pin barrel.

It is also helpful to refer to the printed musical notes for the various chimes. Figure 6 shows the music for Westminster, Whittington, and St. Michael chimes. Even for those who don't read music, there is something to be gained from the printed notes. Look for a pattern of tones going down the musical scale. This is easy to spot on the page and just as easy to see in the rise and fall of the clock hammers as the chime train runs. Westminster, Whittington, and St. Michael chimes all have this pattern at the first quarter, and as the last measure of the third quarter chime.

In some mantel clocks the chime rods do not belong in ascending or descending note order within the clock case, so you cannot synchronize the chime train by watching for the four hammers to fall in sequential order. This adds one step to the task of synchronizing the chimes. Fortunately, the rods will usually be found installed in the correct order, because it is a lot of trouble for someone to unscrew the rods from the gong base and then put them back in a different order. If you have no reason to suspect the rods have been switched, just tap out the pattern of descending notes directly on the rods. This identifies the first quarter Westminster chime pattern you look for as the hammers work.

When the tubular bells are not hanging in the correct order in a hall clock case, your job may be a difficult one. There is no apparent reference point for the repairer who needs to rediscover the correct tube hanging sequence. The problem can be especially serious in an older 9-tube grandfather clock, because of the number of possible ways to hang the tubes. Longest-on-the-left is the most common tube sequence. Some are the opposite, and a few have other patterns. The most complete work on tubular bell melodies and tube hanging sequences is Henry Fried's article, "All About Tubular Chime Clocks" in the April 1982 NAWCC *Bulletin*.

Once the rods or tubes are placed correctly in the case, proceed to synchronize the chimes. First, chime the clock through the hour and then through the first quarter. Next find the chime drive wheel or connecting link between the chime train and the pin barrel. Loosen the set screw holding the wheel, and turn the wheel or the pin barrel by hand to make the hammers operate. Chime movements are not all the same: think about what you are doing and do not loosen any part holding back a mainspring or weight. When in doubt remove all power.

When the hammers rise and fall in the desired pattern, usually notes descending the scale, stop and tighten the set screw again. This will reconnect the pin barrel to the chime train. Chime the movement through an hour or two to be sure of the adjustment. Be certain the last hammer in sequence is never left hanging in the raised position. It is equally important that the first hammer in a sequence does not start to rise when the chime train moves to the warning position. Chime trains are likely to stall if the wheels must start up under a hammer load.

If Westminster is the reference chime used in synchronizing a dual or triple chime clock, it is also necessary to check the other chimes carefully. Westminster, with a four-note measure, has a greater tolerance for error in synchronization than Whittington, which has twice as many notes. That means the last hammer might work fine under Westminster, but another will hang up at the end of a Whittington sequence.

HOW TO ASSEMBLE CHIME MOVEMENTS

Having made the case for taking apart all chime clocks, it is only fair to present a general procedure for putting them back together again. Every model is different, of course, but there is much that is common to all. I recommend that you make sketches to aid in reassembly, but it is not practical to spend hours drawing or taking photographs. A common sense procedure usually gets the job done.

As a first step, sort the various wheels into time, chime, and strike groups. This really isn't hard to do. The time train wheels do not have pins on them, and the escape wheel is certainly easy to find. The pallets, hour and minute wheels, and other easily identified parts will not create a problem. Now separate chime from strike. To some degree, experience will have to teach which is which, or which pieces should be marked before disassembly. Often the chime wheels and fly are larger than the strike, but there are many exceptions. The hammer-lifting star on the strike side is easy enough to locate. But sometimes the two warning wheels are a problem because they are similar.

Sometimes it is best to try fitting wheels in place if you are unsure of their location. Start with the wheels you know, fitting them into one of the movement plates. Continue with the questionable ones, going on the assumption that a wheel turns the pinion above it. Seeing what fits and what doesn't fit usually provides the answers. If a wheel and pinion do not mesh, then they don't belong together. One is fitted backwards, or perhaps it is from another train. I find myself going through this process when I am putting together a chime movement that I have not seen for a long time, or one that I am doing for the first time. It becomes easier with the common movements; if you do three or four of a particular type in a year's time you will remember a lot about it.

Once you have been able to separate the arbors into three groups and know where all the parts should go, you are ready to begin actual assembly. Take a moment, however, to look around for any other arbors which must be added now. A strike hammer arbor or a chime lever may have a pivot on each end or may extend far beyond one of the plates. If so, add the part now. You cannot install it after the plates are together.

Before going on with the assembly, decide which is easier: assembling parts into the front plate or into the back plate. In many movements, the center arbor remains with the front plate, held in place by the cannon pinion. It's the same story with the strike gathering arbor, held by the gathering pallet. Unless you intend to remove the arbors, you will find it easier to assemble the movement by placing all the arbors into the front plate, then adding the back plate. For each movement, there is a best way to set up the job. In the case of the Seth Thomas No. 124 movement, it is better to assemble the arbors into the back plate, which is larger than the partial front plate, and easier to handle. In addition, the center arbor is not held in the front plate by a press-fit

CHAPTER 1 - CHIME BASICS

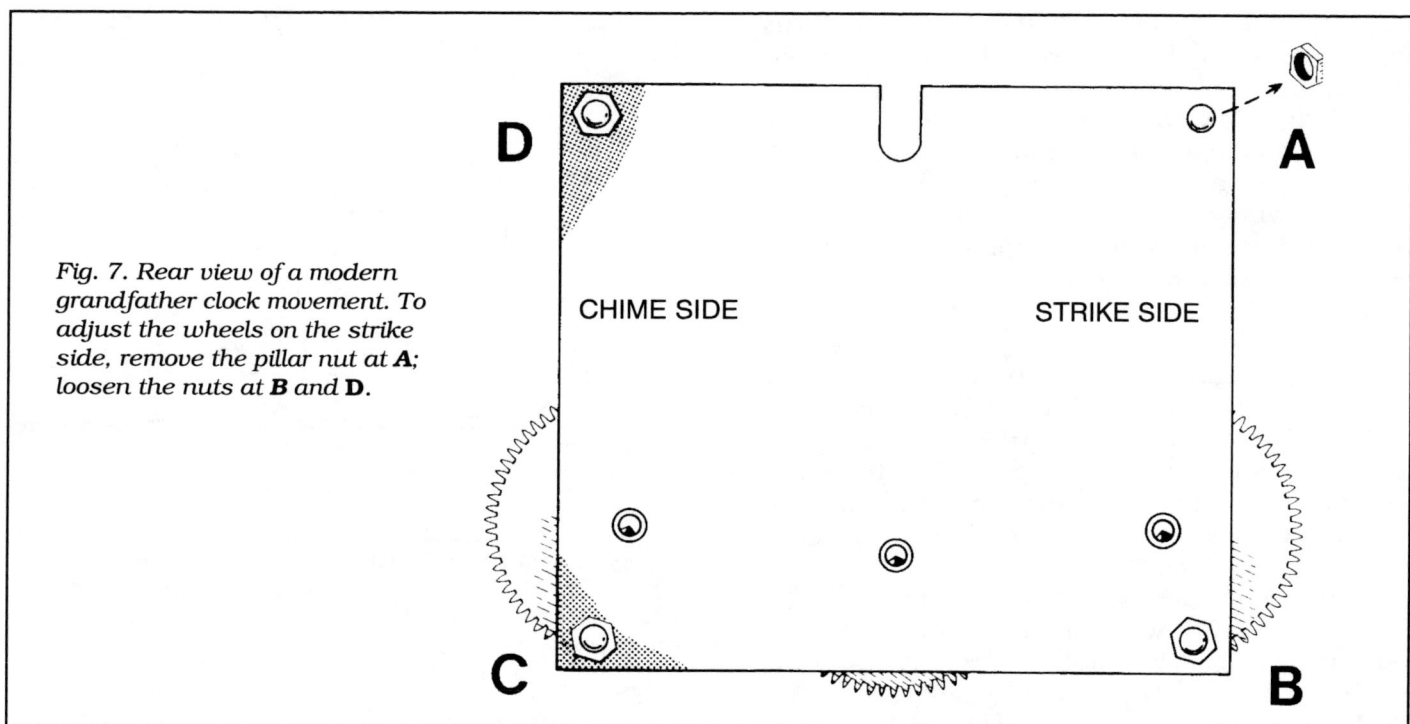

*Fig. 7. Rear view of a modern grandfather clock movement. To adjust the wheels on the strike side, remove the pillar nut at **A**; loosen the nuts at **B** and **D**.*

cannon pinion. It is fastened with a set screw, permitting easy removal. The center arbor can be placed in the rear plate along with the other arbors.

If the chime clock is weight driven, the chain wheels or main wheels with cable drums must go in before the plates are assembled together. Most spring driven chime clocks, plus some of the newer weight driven movements, have barrels or main wheel assemblies which can be installed after the plates are together. In the spring driven clocks, the barrel is to be placed between the plates, then the winding arbor inserted through the front plate. The click wheel is pushed on the arbor and held in position with a cover and set screw. The grandfather movements sometimes have removable pivot brackets fastened in slots in the movement plates.

Even before assembling the plates, some repairers try to synchronize chime and strike wheels for proper adjustment. This is very hard to do. Arbors standing in one plate are not perfectly upright, so it is difficult to see how the wheels will mesh in relation to each other. It's difficult, for example, to set a half revolution of warning run on the strike train this way. During assembly, the parts move about too much. Instead, just concentrate on getting the plates together safely without bending any pivots. This is enough to do in one step, without worrying about adjustments. Put the back plate on, looking for any long pivots which must come through first.

Generally, fit the pivots into their holes going from bottom to top of the movement. As soon as you get the chain wheel pivots, or perhaps the second arbor pivots, in place, add the lower pillar nuts. Screw them on with your fingers, but keep them loose. They are only supposed to maintain the progress made so far. Look for the longer pivots, which are most likely to be holding the plates apart at any given stage of the assembly.

Careful inspection through the top of the movement will usually reveal exactly which pivot is pressing hardest on the plate. This is the next one to ease into the hole. Use tweezers or any small tool that is easy to handle.

When the plates are together, put on all the four pillar nuts finger tight. The nut for the fifth pillar, if there is one, can be left aside until bench adjustments are complete. Turn the wheels in each train to make sure you haven't managed to get an arbor in backwards. It is possible to do just that, without forcing pivots, if the design of the clock will permit certain arbors to go in the wrong way. You may discover that a wheel and pinion do not mesh at all. Or they might bind tightly because they do not match. As soon as everything is between the plates and running smoothly, you are ready to go on.

ADJUSTMENTS

At this point, it is time to make any adjustments which would make it necessary to change the relative positions of chime or strike wheels. Procedures vary with each model, but basically you need to check for chime and strike warning, and for hammer operation. When these are correct, the assembly can be completed. Start with the adjustments on the strike side. Some chime movements (Hermle, for example) do not require you to separate the plates again for adjustment work on the chime side. Strike adjustments, on the other hand, usually involve separating the plates.

Test the strike train by turning the wheels by hand. Install the rack hook, hammers, and the rack, then see how the train operates. If the strike locks with the hammers raised, you must make an adjustment. With many clocks, changing the rest position of the hammers is done just by moving the gathering pallet to a new position on its arbor. But many times you can't get the job done this way. Either the design of the strike train doesn't permit it, or you haven't loosened the gathering pallet. In modern clocks, I leave the strike gathering pallet on throughout the repair, unless I've had to remove it to install a bushing. It is driven on tight, and must be removed carefully. This makes it a time-waster if removed unnecessarily.

Figure 7 (previous page) is a rear view of a chime movement, showing which pillar nuts to loosen when you must change the mesh of strike gears. First, remove the nut at A, on the strike side at the top. If you leave it on, you will probably not be able to separate the plates enough. Loosen the nut at (B), the lower strike side, at least half way. Similarly, loosen nut (D), the upper corner on the chime side. You have to loosen these because the plate isn't all that flexible. However, leave nut (C) finger tight. If there is an additional pillar near the center of the plate, remove the nut. Experience will soon show you how much to loosen the nuts. Try to gently separate the plates with your fingers. If necessary, loosen a nut another half turn to give you the clearance you need to get the pivot out. Or if it appears that many of the pivots are going to pop out, simply tighten down one or more nuts just enough to keep them in place.

The chime train adjustment procedure for loosening the pillar nuts is opposite to that for the strike side. You would remove the nut at (D) and loosen the nuts at (A) and (C).

Figure 8 is a top view of the strike train in a chiming clock, with the plates separated enough for an adjustment. The warning wheel (x) can be changed to a different relationship with the gathering arbor (y) below it in the gear train. One of the fly pivots pops out of the hole, but it is easy enough to slip it back in after the warning arbor is in place again.

Adjustments on some clocks do not involve separating the plates to change the position of any chime wheels. On the Hermle movement the locking is external, so the warning run is set by tightening the locking cam in the right orientation. Herschede tubular bell movements also have an external locking arrangement. But locking is accomplished by the gathering pallet, which can only be installed four

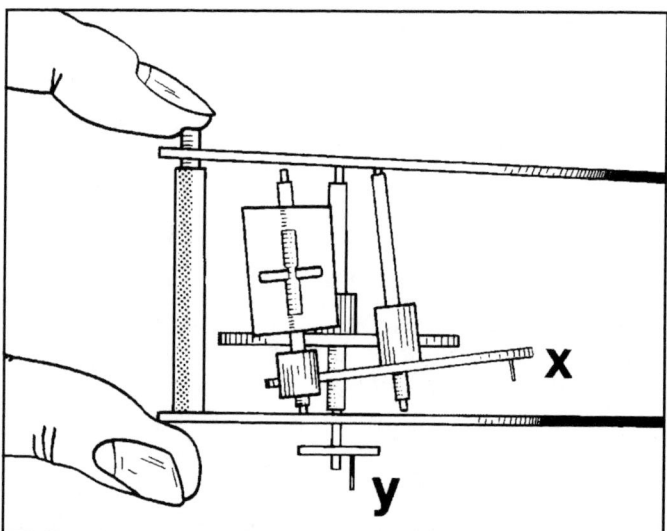

*Fig. 8. Separate the movement plates just enough to ease out one or two pivots. Here the strike warning wheel **x** will be adjusted in relation to the gathering arbor, to set the warning. Note the gathering pallet **y**.*

ways on the arbor. If none of the settings yields the correct warning run, you would have to separate the plates to make the change. An Urgos movement with chime locking between the plates must be adjusted by separating the plates. Each brand of movement is different from the others.

TIPS ON ASSEMBLING CHIME CLOCKS

The most important rule in chime clock assembly is: never force pivots into the holes. This is when they bend or break. If a pivot will not stay in its hole, it usually means another pivot should be seated first. And if you just can't seem to get things together, you should stop for a moment. You've almost certainly got an arbor in backwards or switched from one gear train to another. If you follow this rule, you'll probably never break a pivot. Be willing to take a movement apart again if you must. You may put an entire movement together and then notice something you cannot ignore. It may be a worn pivot hole or a loose click. Once, I began testing a Herschede tubular bell movement, only to find that the maintaining power did not work well enough to keep the escapement working during winding. This was a defect I couldn't live with. I'll admit I was tempted to leave it, but I took the whole movement apart again the next day. Then at least I was satisfied with the restoration.

Remember to synchronize the chime and strike wheels for correct warning and hammer action. To have all these settings right is the hallmark of a good job.

2

COMMON REPAIR PROBLEMS

One of the most fascinating aspects of chime clock repair is the variety. Most chime movements do all the same things—chiming, striking, and keeping time—yet there seems to be an infinite number of ways the mechanisms can be designed to perform these functions. That is the reason this book is based on specific examples of chime movements, rather than general types. The common movements cross the clockmaker's bench many times: hence, the need for a record of each specific model. But before we look at the movements, we need to consider some of the universal repair problems that affect chime clocks.

GENERAL WEAR

Worn pivot holes are the most common defect. Chime trains run with a lot of power, and wear adds up over the years. One or two worn holes will increase friction and ruin performance. Chime speed slows down, and a spring driven train will not run for seven or eight days on a winding; a weight driven train continually fails.

Figure 9 shows an "outside" view of a pivot in a badly worn hole. The pivot is not centered in the oil sink which surrounds it, but is pushed to one side by the wheel below. Often there is a heavy rim of black dirt at the point of wear, indicated by the arrow. By examining all the holes, you should be able to locate several which are clearly much worse than all the others. After you disassemble the movement, concentrate on these. Try to establish some measure of how much wear you can tolerate. It is not practical to install a bushing in every pivot hole. Fortunately, there are few movements which need as many as ten bushings over the three trains. Two or three in a movement are common.

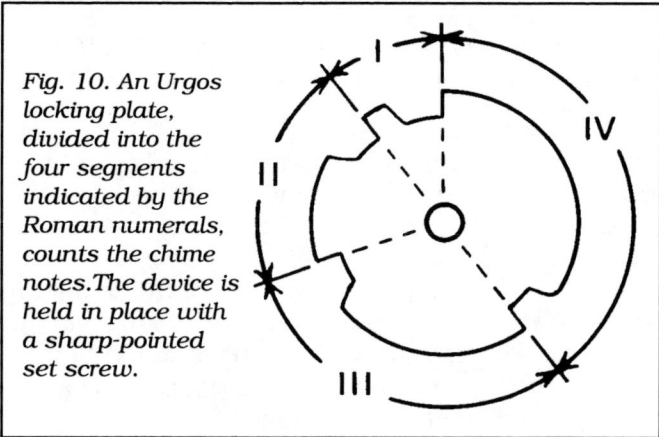

Fig. 10. An Urgos locking plate, divided into the four segments indicated by the Roman numerals, counts the chime notes. The device is held in place with a sharp-pointed set screw.

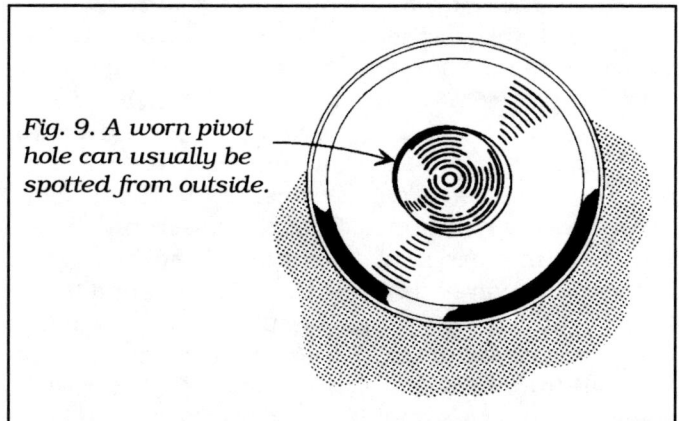

Fig. 9. A worn pivot hole can usually be spotted from outside.

Another problem is the result of over-tightened set screws. Figure 10 shows a locking plate, the counting device for most chime trains. Each of the four segments has a locking slot which must be synchronized exactly with the movement. The locking plate is usually fastened to its long pivot by set screws, some with pointed tips which dig in to prevent slipping. Repairers often turn these screws extremely tight over and over again, as new adjustments are tried. After years of this abuse, the pivot has several raised burrs at or around the desired locking place. As soon as the screw is tightened, it moves into a nearby burr or hole, rotating the locking plate slightly as it does so. With the adjustment lost, the clockmaker tries again with no success.

The same problem occurs with the elongated pivot which carries a chime drive wheel. Figure 11

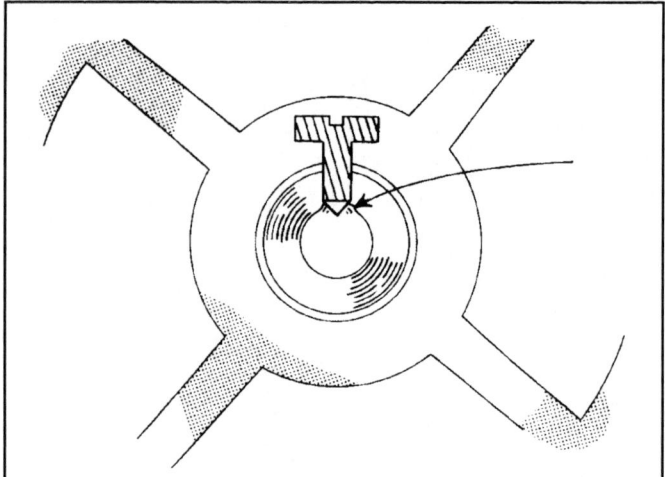

Fig. 11. The set screw which fastens a chime drive gear (or a locking plate, Figure 10) on the arbor. If the screw is over-tightened, the arbor is damaged.

is a cutaway view showing the raised burr. Even when the set screw is backed completely out, it is sometimes hard to pull a chime drive wheel or locking plate over the burr. Twisting and prying can break off the pivot. The burrs can also score the pivot hole during reassembly. To cure the problem, polish off the burr. Then reassemble the movement so the set screws will touch the pivot in a different place. Tighten them with reasonable force. The locking plate and chime drive wheel only need to be tight enough so they will not slip. They do not necessarily have to be so tight that you cannot twist them by hand. Especially on a large wheel, you have tremendous leverage as you twist it: in actual operation the movement does not see forces like this.

Still another common defect is the worn or rough warning lever. The acting surface is indicated by the arrow in Figure 12. If the surface is rough, the lever may not always release the warning pin at each

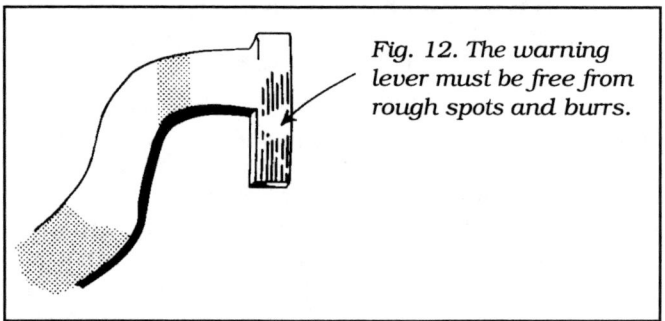

Fig. 12. The warning lever must be free from rough spots and burrs.

quarter hour. If this happens, the clock does not move from the warning position to begin chiming. It can be puzzling because the failure is intermittent. The clock usually chimes when the hands are turned to a chime point, but soon fails running on its own. When the clock is apart, polish the warning lever with strokes in the same direction the pin slides, not across. Polishing is easy to do at this point in the repair, but may be impossible after the clock has been reassembled. Test the clock to make sure the warning pin hits the lever safely at the warning point, then slides smoothly off at the instant chiming is to begin.

HAMMER ASSEMBLY

The hammer assembly in any chime clock is also the source of many problems. The assembly contains many separate parts subject to wear, incorrect assembly, and substitution with wrong parts. Figure 13 shows a hammer assembly from a Seth Thomas No. 124 movement. The hammer arm rubs on a brass pillar as the hammer falls, to dampen the bouncing. Always lubricate at the point indicated by the arrow to help prevent jamming. Clocks will fail during testing or come back later, because the chime does not work. A thorough check may turn up nothing more than a binding action as the

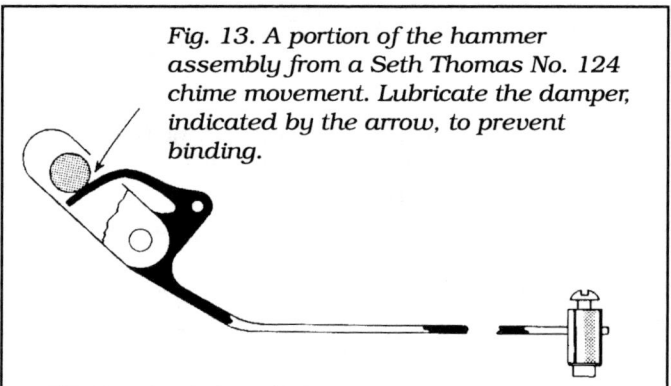

Fig. 13. A portion of the hammer assembly from a Seth Thomas No. 124 chime movement. Lubricate the damper, indicated by the arrow, to prevent binding.

hammer arm contacts the brass pillar. You may have to bend the arm slightly to keep it from rubbing too hard against the pillar, then lubricate the contact surfaces with clock grease.

In older chime clocks, there are washers on the hammer shaft, between each hammer. Figure 14 indicates two of the washers. They act as spacers from one hammer to the next, to keep them from fouling each other. Three or more washers may be found over the rear hammer, under the bracket plate for the assembly. These may be of several thicknesses, some brass and others steel.

When you remove the bracket plate, the washers may drop off and roll away. You should try to get them all back in the right sequence after cleaning. But do not assume the washers in the assembly are the original ones. Test the assembly to see how the hammers move. If there is excessive wandering of the hammers along the hammer shaft, then washers are certainly missing. If you have a washer between each hammer, perhaps all you are missing is an additional washer to be installed before the bracket plate goes on. If, on the other hand, the

CHAPTER 2 - COMMON REPAIR PROBLEMS

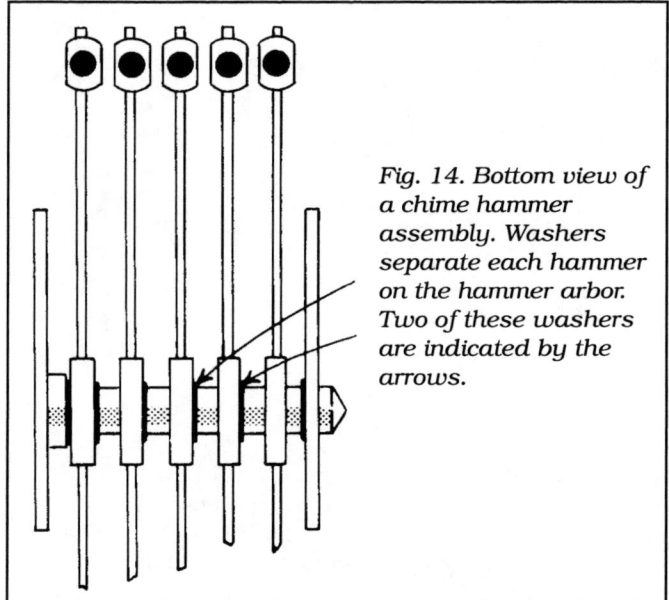

Fig. 14. Bottom view of a chime hammer assembly. Washers separate each hammer on the hammer arbor. Two of these washers are indicated by the arrows.

hammers are tight, you may have too many washers. As a final check, make sure the pins on the pin barrel line up with the hammer tails. You may have to shift some of the washers from one place to another to restore the correct hammer positions.

Hammer heads always deserve careful attention. The first thing a customer will notice, after he or she has spent money to have the clock repaired, is the chime sound. Modern clocks with plastic inserts for hammer heads do not present a problem. But older clocks usually have leather hammer heads, and these require replacement from time to time. You probably will not want to replace hammer heads unless it is really necessary. Look carefully, however, at each one. You may find considerable variation among them. Figure 15 shows adjacent hammers from a chime clock. The hammer on the left is good, and may be an earlier replacement. But its neighbor is worn down on one side, and the leather is hard and old, or perhaps soft and oil soaked. Even though the right-hand hammer may make just as loud a sound as the other one, it will be different in tone quality. The customer may object to it. Replace any questionable hammer, or all of them if it is necessary.

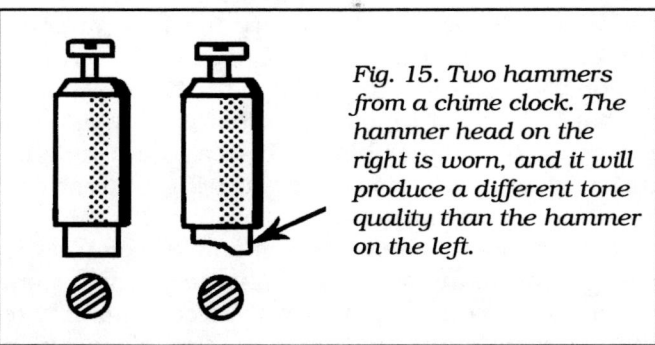

Fig. 15. Two hammers from a chime clock. The hammer head on the right is worn, and it will produce a different tone quality than the hammer on the left.

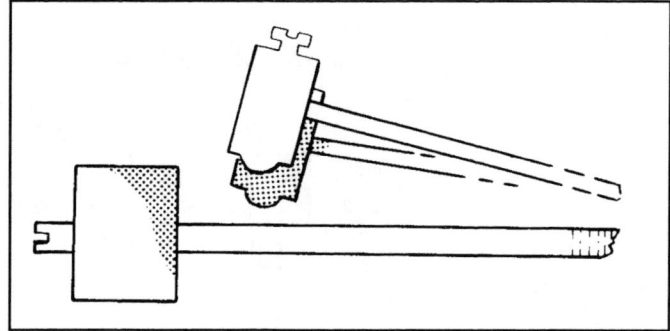

Fig. 16. Hammer clearance. The shaded hammer will hit its rod, but the white hammer will not make a sound.

Hammer clearances to the rods must be correct and uniform. If they are too close, the hammer bounces several times on the rod with each hammer blow. If too far away, the sound is faint or not there at all. The correct clearance is usually between 1/16" and 1/8". Figure 16 shows two hammers in a side view. The front hammer cannot hit the rod because it is set too far away from it. Even if both hammers hit their rods, they would not be equal in loudness or tone quality.

Figure 17 illustrates a set of four hammers. The alignment, or side spacing, of the hammers is incorrect. The hammer at **a** is bent so far to the left that the hammer head will miss the rod. Instead, the brass part of the hammer will make a sharp, metallic sound as it glances off the steel rod. The hammer at **b** is straight, but it is too close to the hammer at **c**. Hammer **c** should be bent to the right. Hammer **d** is correct.

To adjust a hammer shaft, grip it below the point of bending, with pliers. Then bend with fingers or another pair of pliers. A pair of slotted levers can

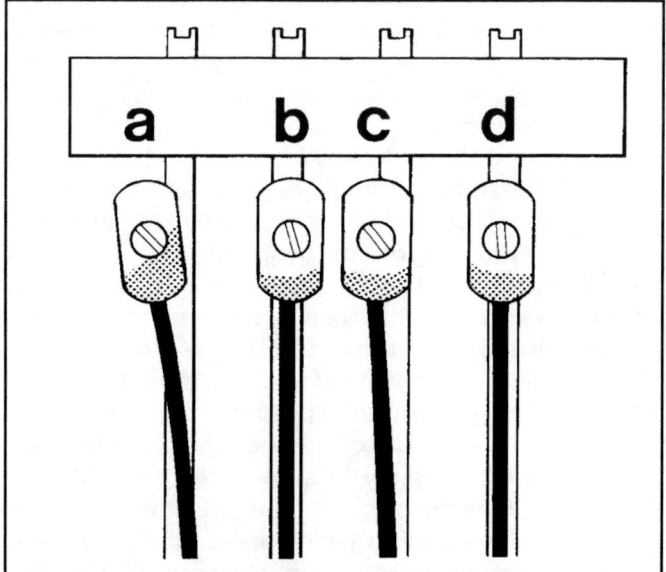

Fig. 17. Hammer alignment. Hammer **a** misses the rod; **b** is correct, but **c** will vibrate against it; **d** is correct.

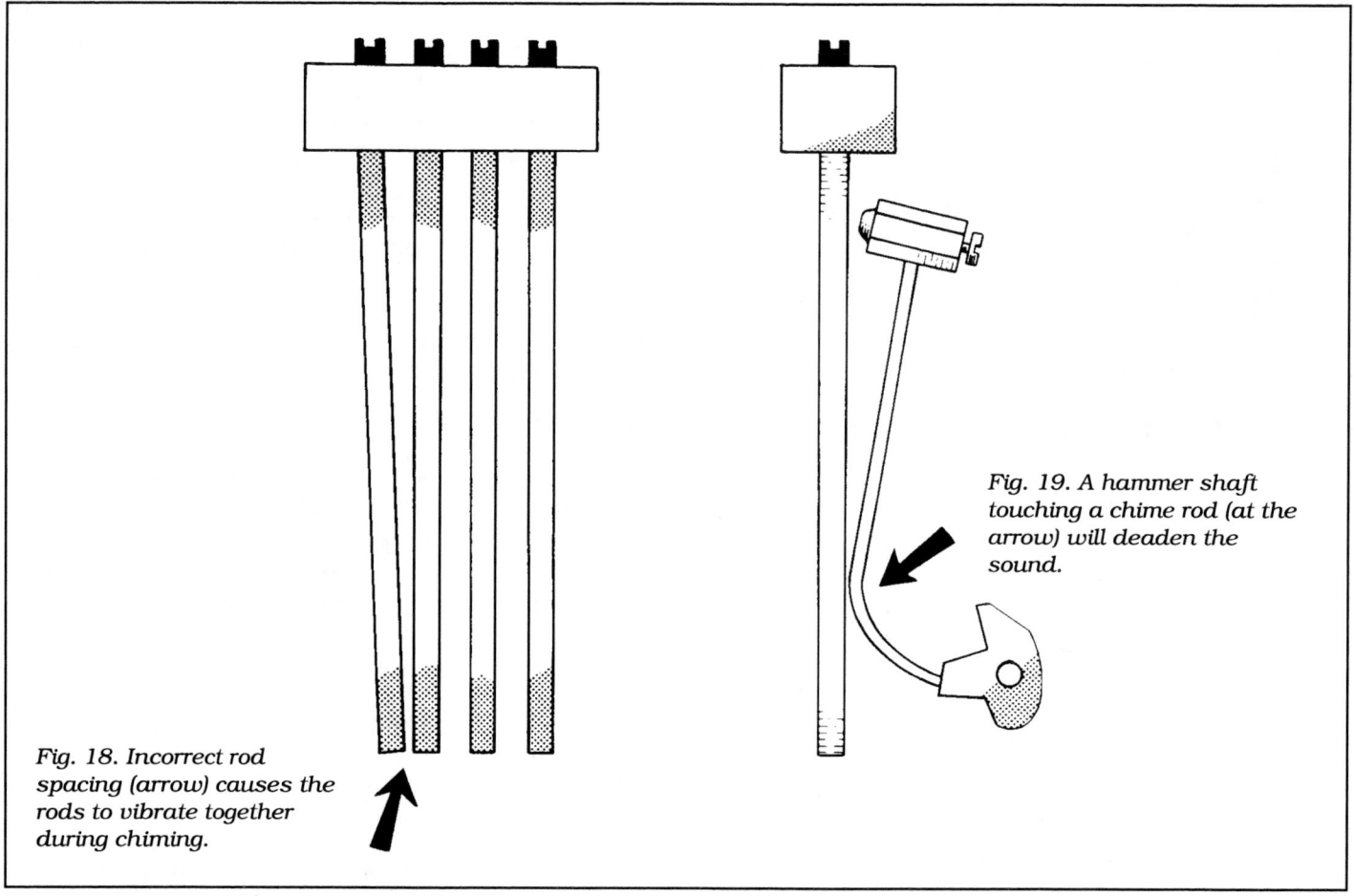

Fig. 18. Incorrect rod spacing (arrow) causes the rods to vibrate together during chiming.

Fig. 19. A hammer shaft touching a chime rod (at the arrow) will deaden the sound.

also be made to make the adjustment. In some cases you could just as easily bend with the fingers alone, but sooner or later you will get into trouble. If you do not support the hammer shaft firmly, much of the bending force is exerted at the base, where it is fastened to the hammer lever. You may twist the shaft loose, making it impossible to get it to stay where you put it. That is bad enough at the bench, but a very unpleasant situation for a service visit in the customer's home.

Hammers will usually vibrate together if the alignment of the heads is not correct. Some clocks allow only a small clearance because of close rod spacing or large hammer heads. In these, alignment is critical. Others permit more of a tolerance for error: the hammers do not vibrate together unless they are so far out of line that some of them miss the rods.

Figure 18 shows a condition often found in grandfather clocks. Two rods hang down quite close to each other. They vibrate together during chiming, creating discordant sounds. Sometimes the rods clash when someone walks past the clock. On the strike side the problem is the same, with the rods vibrating together especially on long counts—the last few strokes at 10, 11, and 12 o'clock. The solution is to grasp the offending rod near the bottom, and pull it until it hangs straight. Unfortunately, there is a certain amount of risk. Up inside the gong base, the steel chime rod is turned to a small diameter where it fits tightly into the brass mounting screw. The bending effort takes effect here, not on the lower part of the rod. In actual practice, you have to take the risk of breaking off the rod if you must repair the problem in the customer's home. Just don't bend any more than necessary to keep the rods from touching each other.

Chime sound is completely deadened when hammer shafts touch the rods. Certain models have inadequate room for the two rows of hammers and rods. The hammer shafts either extend between rods or rest too close to them. Figure 19 shows the point of contact. Sometimes it is very hard to pinpoint this condition as the cause of poor chime sound. The only cure for it is to re-shape the hammer shaft so that it does not touch the rod or vibrate against it with each hammer blow. This can be a tedious and frustrating task to perform during a house call repair on a grandfather clock.

Some modern chime movements develop early wear on the acting part of each hammer lever, from the brass pins on the pin barrel. A weak chime note often attracts your attention in the first place. Upon

CHAPTER 2 - COMMON REPAIR PROBLEMS

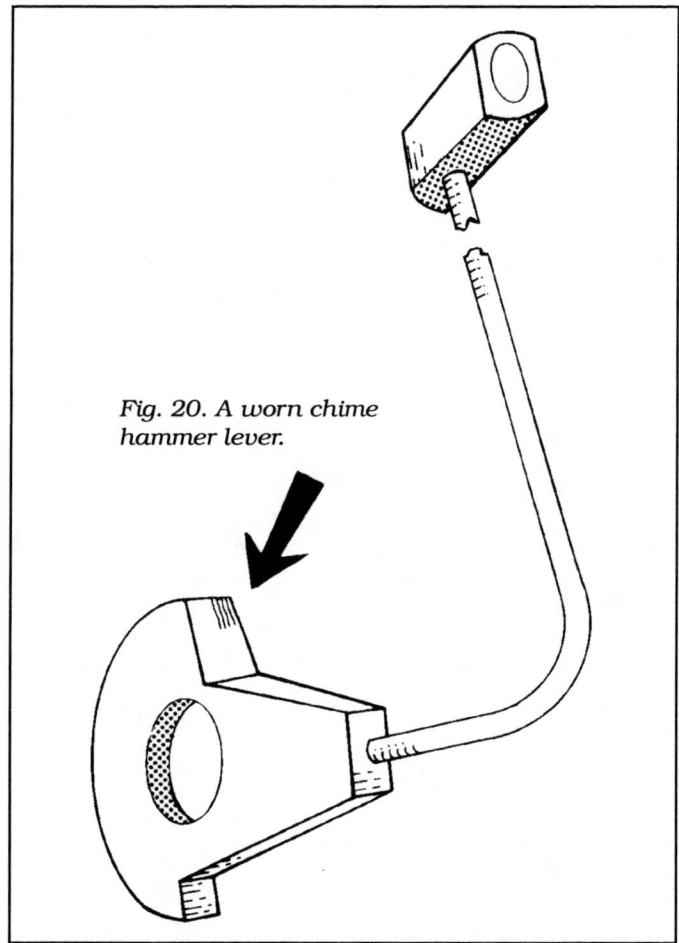

Fig. 20. A worn chime hammer lever.

investigation it seems that one or two of the chime hammers do not lift as high as the others. Severe wear of the hammer lever may be the cause. Figure 20 shows a single hammer lever, with an arrow drawn to indicate the worn portion. All hammer levers will have a normal rubbed appearance from the thousands of hammer strokes performed. But some develop deep grooves, particularly at one side. In a Westminster chime movement, a missing spacer washer may allow the hammer levers to wander. Once the pin begins to wear a groove at the side of the lever, it just gets worse. Besides doing something about the wear, it is necessary for you to add one or more washers to eliminate the extra play in the hammer assembly.

The hammer levers in a triple chime clock may be worn on the side for a different reason. The triple chime shift mechanism may be the cause. If it does not move the pin barrel to the correct places in relation to the hammer assembly, one or more hammer levers will wear out. As a start, the shift mechanism should be adjusted.

Worn hammer levers in a modern movement are easy to correct by replacing the entire hammer assembly. If the assembly is not available, notch out the worn spot in the hammer lever and fit in a new portion. After you solder it neatly in place, file and polish the new piece until it conforms to the original size. By comparing your work to the unworn hammer levers, you can make a good repair. To check the result, watch the hammers move. If the repaired hammer lifts to the same height as the others, the repair is successful.

SPRING DRIVEN MOVEMENTS

Repairers open up many chime movements to find extensive damage from broken springs. The movements rarely get the benefit of disassembly and overhaul anyway, but it seems that the springs suffer the most from neglect. In turn, these springs break and damage the movement. Some mainsprings break without warning, but many let go as a result of torn outer ends which would have been detected if the spring had been removed from the barrel for cleaning and inspection. Broken pivots and teeth are common in the gear train affected by the bad spring. The springs are so powerful that the damage may include torn-out barrel teeth, a broken pivot or pinion on the second arbor (next to the barrel) and broken teeth further up in the train.

You may find that a chime movement has been cleaned while fully assembled, spring barrels and all. Most barrels are almost completely sealed, so any solvent which seeps inside during the cleaning remains trapped there. Gradually, the mixture of solvent and dirty oil thickens into a real mess. Before you clean the movement, remove the barrel covers and take out the springs. Just by restoring the springs to a clean, lubricated condition, you will greatly improve the performance of the clock.

Always use a mainspring winder to remove and insert springs. It's safer than trying to wind by hand, and it avoids distortion of the spring. Replace springs if the ends are torn. Give thorough attention to all winding parts, including clicks, click wheels, and click springs. Don't take chances on a later failure.

GEARED WINDING MECHANISM

Figure 21 illustrates the winding mechanisms from an older Herschede mantel clock, manufactured when the firm was still in Cincinnati. Seth Thomas and Ansonia also produced the same type of heavy chime movements for mantel clocks in the '20's and '30's. Many of the cases were the familiar tambours. The Herschede mantel clock movement looks like a smaller cousin of the large tubular bell chime movement. It sits on a seatboard inside the mantel clock case, with additional brackets on the upper pillars. Lacking in equipment, many a repairer will ignore these strong mainsprings entirely. Because the springs are too heavy to wind into the barrels by hand, he just pops off the barrel cover and adds

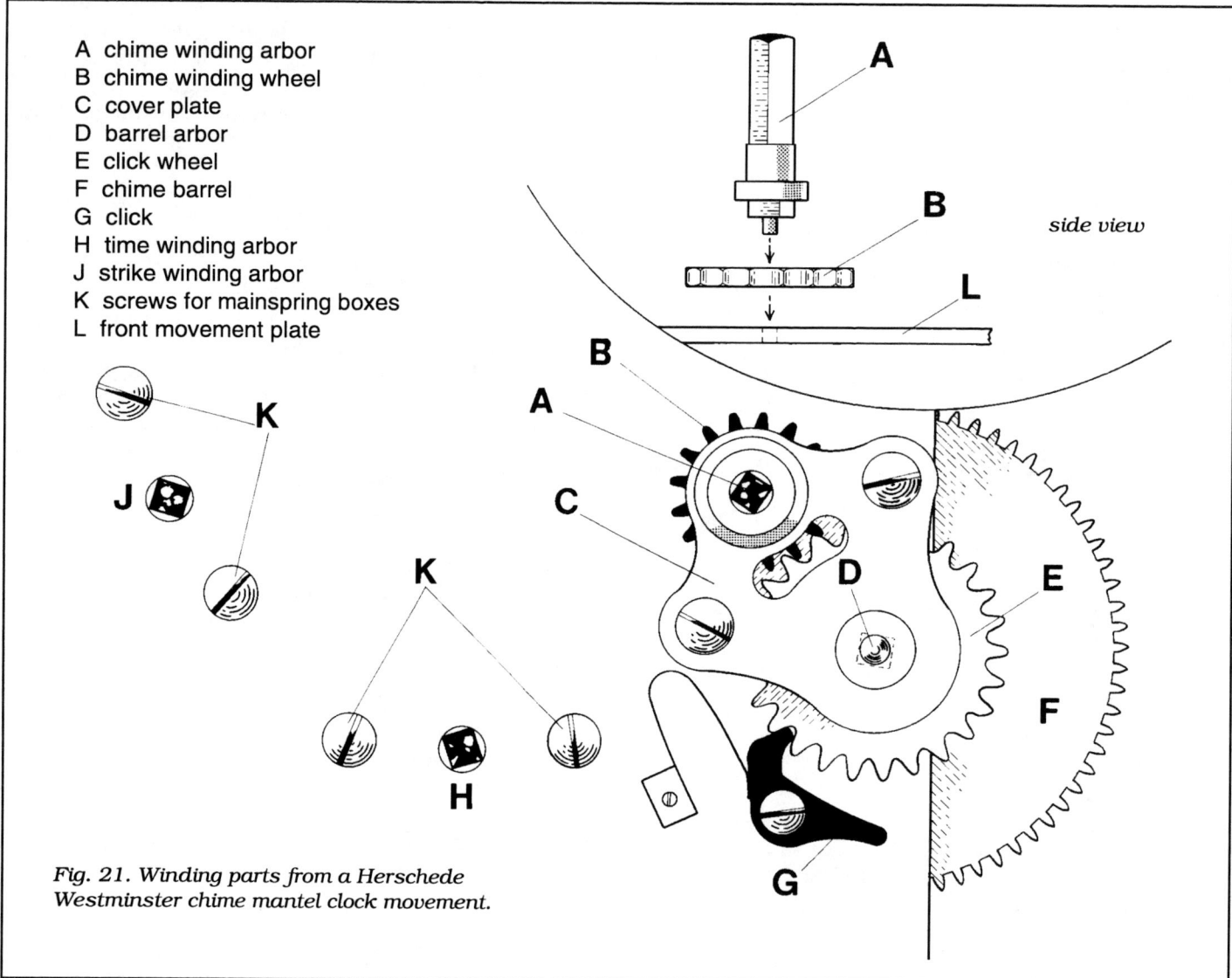

A chime winding arbor
B chime winding wheel
C cover plate
D barrel arbor
E click wheel
F chime barrel
G click
H time winding arbor
J strike winding arbor
K screws for mainspring boxes
L front movement plate

Fig. 21. Winding parts from a Herschede Westminster chime mantel clock movement.

more oil. After many years, the springs are so dirty that they cannot provide enough power for the clock.

The components of the Herschede geared winding mechanism are shown in Figure 21, parts A-F. The inset (top) shows a side view of the winding arbor and winding wheel. The click wheel (E) has more teeth than the winding wheel (B), thereby giving the clock owner a mechanical advantage of about 1.6 to 1 on winding. The springs, especially the chime, are so strong that it is extremely difficult to wind them otherwise. The geared winding mechanism also relocates the chime winding arbor an inch closer to the center of the movement. This brings the keyhole within the area of the five-inch dial.

Geared winding mechanisms eventually wear out. The teeth on the winding wheel and click wheel become badly worn as the owner turns the key against the spring pressure. Increasingly, the teeth butt each other. Winding becomes rough as it becomes necessary to exert greater pressure on the key. It becomes harder to get the "feel" of the spring—to know when it is nearly wound up. The clock owner ends up winding the mainspring only halfway, or winding to the end without getting any warning to ease up on the key. This is partly the fault of the two-gear winding design, but it is certainly worsened by worn teeth. Clicks often fail from the pressure, allowing the mainspring to let down suddenly. Damage is certain. Many of the Seth Thomas No. 113 models had double clicks which provided a back-up in case one click or click spring failed. Then there is always the hapless amateur who removes the cover plate from the winding mechanism because he doesn't know what it is. The spring lets go explosively.

MAINSPRING BOX

The Herschede clock has another type of mainspring setup in addition to the geared mechanism just described. The time and strike springs are housed in

CHAPTER 2 - COMMON REPAIR PROBLEMS

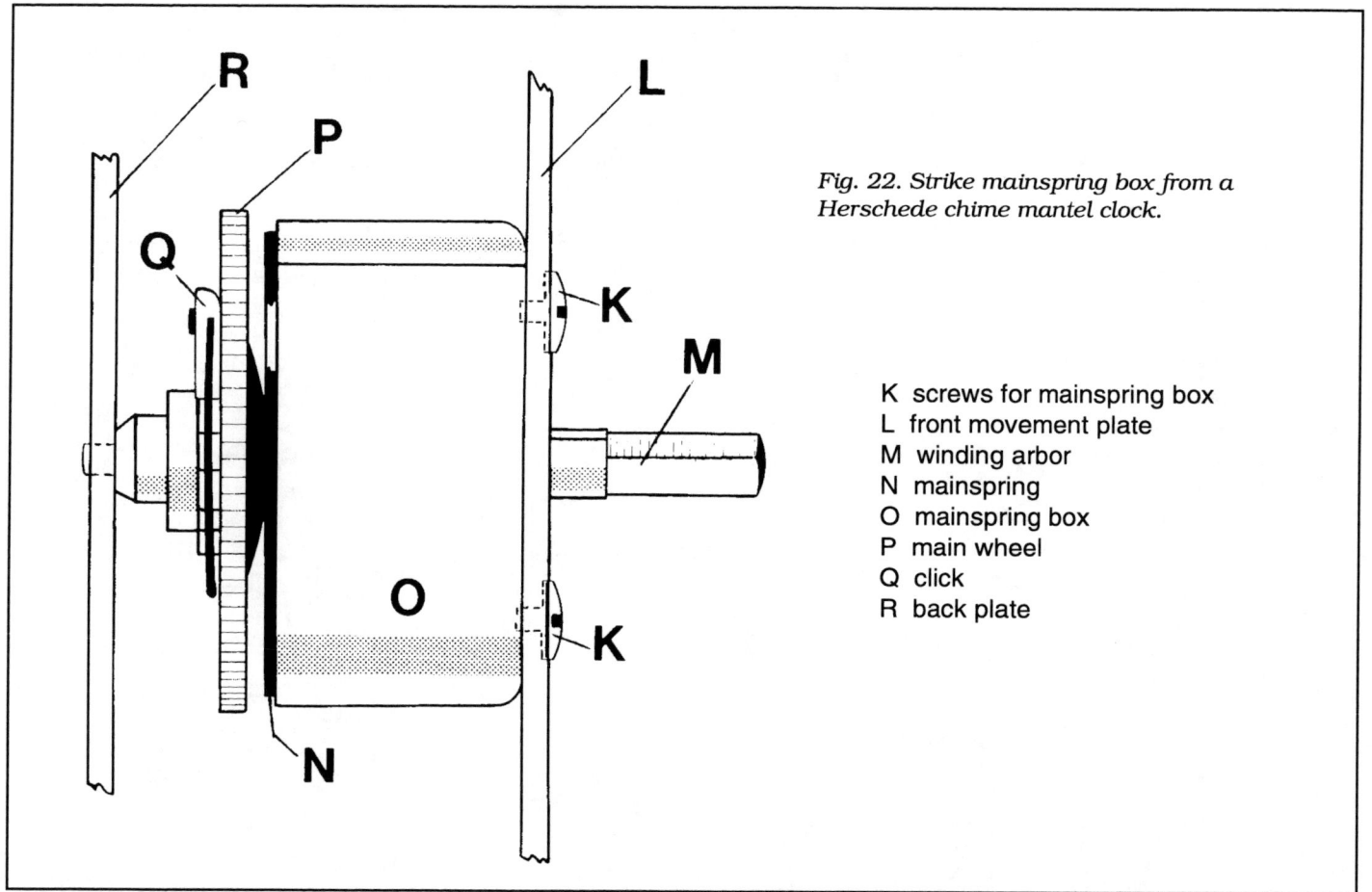

Fig. 22. Strike mainspring box from a Herschede chime mantel clock.

K screws for mainspring box
L front movement plate
M winding arbor
N mainspring
O mainspring box
P main wheel
Q click
R back plate

mainspring "boxes" which are not rotating barrels. They serve only to contain the springs as they expand, and perhaps to keep them cleaner than an open mainspring. Figure 21 shows the locations of screws holding the time and strike mainspring boxes to the front plate. In Figure 22, the main parts of this mechanism are drawn. The Seth Thomas No. 124 features this design for all three springs.

The "boxed" spring and main wheel units are hard to service. Since they wind from the closed end of the spring box, they are hard to accommodate in most mainspring winders without some modification. In the Seth Thomas No. 124 the boxes are often riveted instead of screwed to the plate. This makes them even harder to handle. The clicks and click springs are attached to the main wheels and are surprisingly light in construction. Before you can tighten a rivet, you must remove the wheel and arbor from the spring. The click spring is not especially durable, either. Most are flat springs, and once pulled away from the wheel it is next to impossible to get them back.

WEIGHTS AND CABLES

One of the biggest changes in the grandfather clock industry since the 1970's has been the return of the cable driven clock. Up until that time, the old grandfather clocks were cable drive, the new ones chain drive, with few exceptions. A kinked or frayed cable will cause a modern clock to stop because the cable will not wind smoothly on the grooved winding drum. The cable crosses over itself, cancelling out most of the effect of the weight. Cable guards, usually made of plastic, are installed to help prevent this. The clear guards wrap around the cable drum to hold the cable in the grooves when the weights are removed. The small diameter brass cables used in these clocks also help, because they are not as springy as the thicker cables in older clocks.

Figure 23 shows a Kieninger main wheel. The plastic guard (4) slips over the winding drum (3), encasing the drum. The guard is held in place by a screw (9) which ensures that the guard remains stationary as the drum rotates during winding or running. The cable (1) fits in the keyhole shaped notch in the drum. To eliminate the need for a knot in the end of the cable, the end is made from a solid piece of brass, solder-dipped to prevent fraying. If it is necessary to release the click (8) to let out the cable, it is easy to reach and depress the click. Don't forget to remove the weight first. The Kieninger design

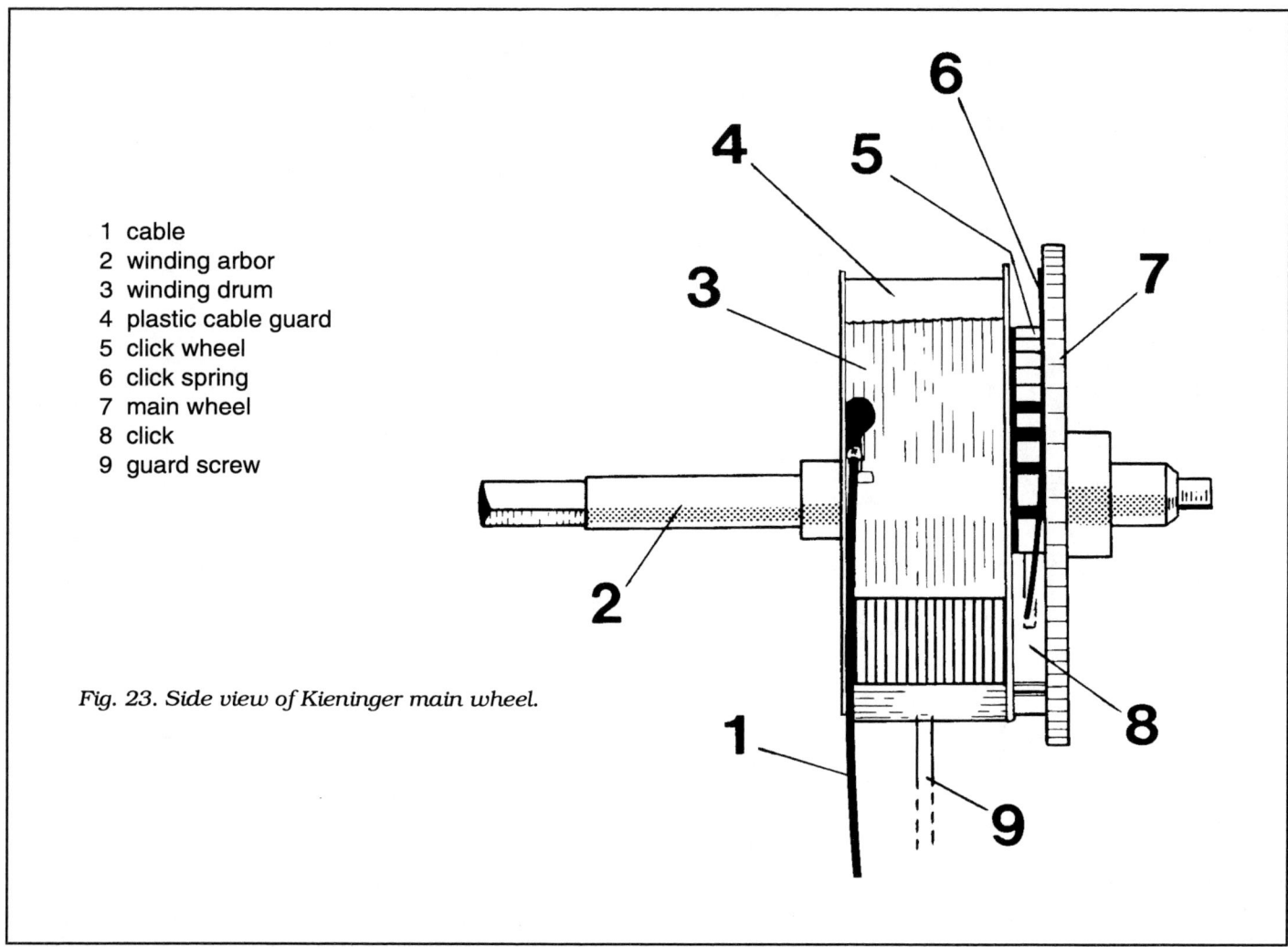

1 cable
2 winding arbor
3 winding drum
4 plastic cable guard
5 click wheel
6 click spring
7 main wheel
8 click
9 guard screw

Fig. 23. Side view of Kieninger main wheel.

shown is one of many produced by the manufacturers. Some are better than others because of differences in the guards and cables. Main wheels with double clicks are especially difficult to service because it is so hard to reach both at the same time.

If you find a modern cable driven clock mysteriously stopping or failing to chime, remove the weights and let out the cables all the way. You may find the cable was crossed over itself. In other instances, the end has come out of the slot in the drum, allowing the brass cable end to bind on the inner surface of the plastic cable guard. The weight usually doesn't fall because of the many wraps of cable around the drum, yet the binding stops the movement.

Always replace the cables if they are not perfect. If a clock is more than five years old or has been worked on before, you may find it necessary to replace them. Once the cables become twisted from handling, kinks develop. The cable jumps out of the groove at these kinked spots, and there is nothing you can do to prevent it. The cable begins to go all over the drum during winding. Fraying is also a problem. The cables are thin, and flexing breaks them down. The strands break most often at the top end, where the cable bends to enter the drum. A falling weight will result in a badly dented shell and probably a broken case bottom. Never try to "get by" with old cables.

CHAPTER 2 - COMMON REPAIR PROBLEMS

CHECKING THE CABLE ENDS

Whenever you remove a weight, check the seatboard cable end before hanging the weight on again. Sometimes the cable end will pop out of the bracket. The usual bracket is a slot or keyhole arrangement which keeps the cable seated as long as the weight is there to pull down on it. Consider taping the slot closed after the cable end is inserted. As a safety feature, Kieninger developed the cable bracket shown in Figure 24. It is installed between the movement plates in small cutouts. If the bracket is lifted and angled into the horizontal part of the slot, it can be swung to the side until it is even with the bottom of the movement. This keeps it flat and out of the way when the movement is on the bench. The cable end is inserted in the round keyhole slot shown in the drawing. Once seated in the bottom of the bracket, the cable end cannot come out when the weight is removed. It takes a deliberate motion to pass the end through the keyhole slot again.

Instead of having special brackets, many clocks have the end of the cable pushed through a hole in the seatboard and tied off with a knot. First of all, it is helpful to heat the end of the cable and flow some soft solder onto it to prevent the end from fraying. Then a knot such as the one shown in Fig-

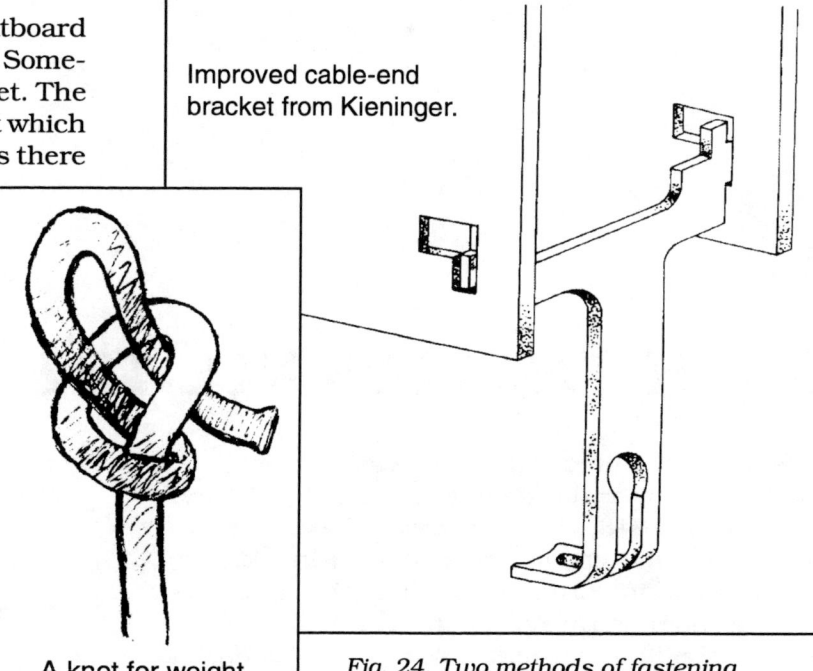

Improved cable-end bracket from Kieninger.

A knot for weight cords or cables.

Fig. 24. Two methods of fastening the ends of clock cables.

ure 24 gives an attractive appearance to the cable end. The loop can be made any size. The downward force of the weight tightens the grip on the loop instead of tending to pull it through the hole in the seatboard.

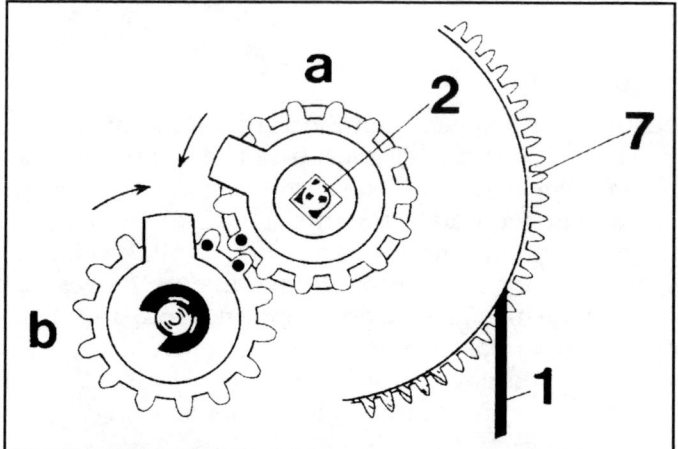

Fig. 25. Stop-works mechanism for a Kieninger cable-driven movement.
1 cable
2 winding arbor
7 main wheel
a stop-works gear mounted on winding arbor
b stop-works gear mounted on idler post

STOP-WORK MECHANISMS

Many of the newer cable drive clocks have stopworks on the winding arbors. These are beneficial to the owner and the repairer because they prevent over-winding and jamming of weights. Figure 25 shows a stop-work setup from a Kieninger KSU movement. At the end of the winding travel, the gears come together, butting long tooth against long tooth. The same happens at the lower end of the weight fall.

If you have replaced a cable or put a movement together, you will need to reset the stop-works. Some repairers have actually removed the mechanism rather than try to set it, but it is not difficult to do. First remove the gear (b). Wind the weight almost as high as you feel it should go, allowing some clearance between the pulley and the seatboard. Continue by replacing (b) and arranging both gears (a) and (b) so the long teeth will contact each other as soon as further winding is attempted.

3

SETH THOMAS NO. 124

Seth Thomas produced an impressive line of clock movements over the course of its long history. Founded in 1813, the Connecticut-based firm finally ended movement production in the 1950's, after it began importing German movements. Fortunately, many of the American-made Seth Thomas clocks are still in existence. The No. 124 is one of the most plentiful.

Introduced about 1924, the No. 124 was manufactured over a 30-year period. It was used in a variety of tambour, arched-top, and bracket style mantel clocks. The 1926 Seth Thomas catalog included nine mantel clocks with the No. 124 movement, ranging in price from $45.00 to $60.00. Figures 26 and 27 show two of the clocks from the 1920's. The No. 124 movement is pictured in Figure 28. Figures 29, 30, and 31 are drawings of the overall movement. Use these and the detail views in Figures 32 through 38 to help you identify the parts of the movement.

The No. 124 is a pendulum movement powered by three mainsprings. At first glance, the springs appear to be enclosed in barrels, but these are not barrels at all. They are mainspring boxes as described in Chapter 2. In some of the No. 124 movements, the mainspring boxes are riveted to the lower front movement plate, and in others they are held by screws. The "boxes" limit the movement of the mainsprings, which would otherwise interfere with each other as they unwind and expand. The main or great wheels are separate from the mainspring boxes: the wheels rotate as they drive the gear trains. The mainspring boxes are stationary.

You must still remove, inspect, and clean these hole-end mainsprings, just as you would do if they were fully enclosed in barrels. Admittedly, it is difficult to get these springs in and out of the mainspring boxes. Even with a mainspring winder, it is very hard to handle them when the boxes are riveted to the plate. Experience shows that many of the mainsprings have cracked or torn ends. For this

Fig. 26. Seth Thomas Chime 90 mahogany tambour clock. Base 21-3/4 inches wide. From the 1928 catalog. Image courtesy Seth Thomas Div. of General Time Corporation.

Fig. 27. Chime 96 clock with No. 124 chime movement. Mahogany case 10-3/8 inches high with inlaid marquetry. From the 1928 catalog. Image courtesy Seth Thomas Div. of General Time Corporation.

reason, it is especially important to examine the springs. If a bad spring goes unnoticed, it may soon fail. Replace a spring if the end is damaged, or if the spring has been distorted into a conical shape.

Replacement mainsprings are available from clock suppliers. There may have been changes to the size springs used over the manufacturing history of the No. 124, but the following currently available sizes will power the clock:

	Width	Thickness	Length
Time	11/16"	.014"	54"
Strike	11/16"	.014"	54"
Chime	3/4"	.015"	72"

Westminster chimes are sounded on four rods, with the hour struck on the fifth rod. The rods are mounted horizontally on the floor of the case in tambour models. Bracket style and arched-top cases are narrower, so the rods are mounted diagonally

CHAPTER 3 - SETH THOMAS NO. 124

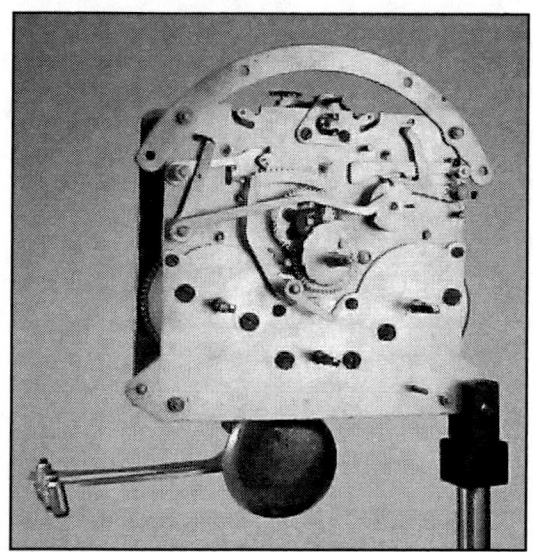

Fig. 28. Seth Thomas No. 124 Westminster chime movement. Dimensions: Height 8 inches, width 6-3/4 inches, depth 5-1/2 inches.

to conserve space and yet permit the use of the same length rods. The hammers in Figure 31 are for the tambour clock case with horizontal chime rods. Figure 32 shows the alternate hammer arrangement for clocks with diagonal rods.

Some No. 124 movements have a silencer arbor as illustrated in Figure 33, turned by the small end of the winding key inserted through an opening in the dial near the numeral 4. Turn it counterclockwise to silence, back again to chime. Many owners have broken off the arbor by trying to force it in a complete circle; check the arbor whenever you estimate repair charges on a No. 124. It is important to instruct the customer on how to silence the clock without damaging it.

In other No. 124 clocks (the No. 124 E is an example), there is no silencer arbor. To silence the chimes, open the back door of the clock and locate the brass tab, 14 mm long, on the silencer lever. The words "chime" and "silent" are stamped on the chime bracket plate. Push to the right for silent, left for chime. This setup operates the same way as the through-the-dial silencer, but without the convenience (or the problems) of the latter. When activated, the silencer lever pushes up on the chime hammer levers.

DISASSEMBLY

Disassembly and cleaning are a vital part of the repair job, often skipped by those who want to avoid hours of work. For this reason, most of the No. 124 movements brought for repair are long overdue for a thorough overhaul. It is almost a foregone conclusion that bushings will be needed for several of the pivot holes. Examine the movement before you take it apart. You could spot a defect you might miss later as you work with the individual parts.

The first step in disassembly is to let down the three mainsprings. Use a let-down key, of course. It may be difficult to lift the small clicks, mounted on the main wheels facing the rear of the movement. A hook made of stiff wire is a good tool for reaching under the clicks. Do not bend the click springs away from the wheels, because you may not be able to bend them back again. After you let down the springs, remove the lower front movement plate with spring boxes and main wheels attached. Do not lose the spacer washers on the two lower movement pillars. These are absolutely necessary for proper endshake of the winding arbors. The lower movement plate overlaps the upper, and the spacers act to keep the two plates parallel.

Proceed to remove the parts in an orderly fashion. Remove the suspension unit and regulator arbor. After you take off the front pivot plate for the pallet arbor, carefully work the crutch and pallets out. Refer to Figure 29 to identify the front movement parts as you remove them. Take off the strike lifting lever (5), hour wheel (15), and minute wheel (14). The cannon pinion is held on the center arbor by a small set screw. Remove the cannon pinion carefully.

Unhook the rack spring, then take off the rack (13). If you take care in using tweezers to unhook the rack spring from the post, you may be able to re-use the spring. Do not retain a damaged rack spring. Remove the rack hook (2) and the gathering pallet (7). Be careful not to bend or break the long pivot which carries the gathering pallet. Unhook the wire spring from the post near the chime lock piece. Loosen the set screw and remove the locking plate (11), chime lock piece (10), and chime correction arm (9).

Turning to the rear of the movement, refer to Figures 31, 32, and 33. First, remove the entire hammer assembly. Begin by removing the chime bracket plate (28). If you have the hammer arrangement shown in Figure 31, you must also take off the small oval plate, held by one nut, which holds the lower hammer assembly in place. Remove the pin barrel (27) and silencer lever (51). Between each of the hammer levers (33 and 44) you will find a thin steel washer, and a thicker brass spacer washer immediately under the chime bracket plate. Do not lose these washers, which are necessary for the proper fit of the hammer levers in the assembly. The hammers must not be tight, but they should not wander forward and back during operation, either. If the movement is equipped with the silencer arbor as shown in Figure 33, remove the assembly. Finish removal of the rear movement parts by taking

CHIME CLOCK REPAIR

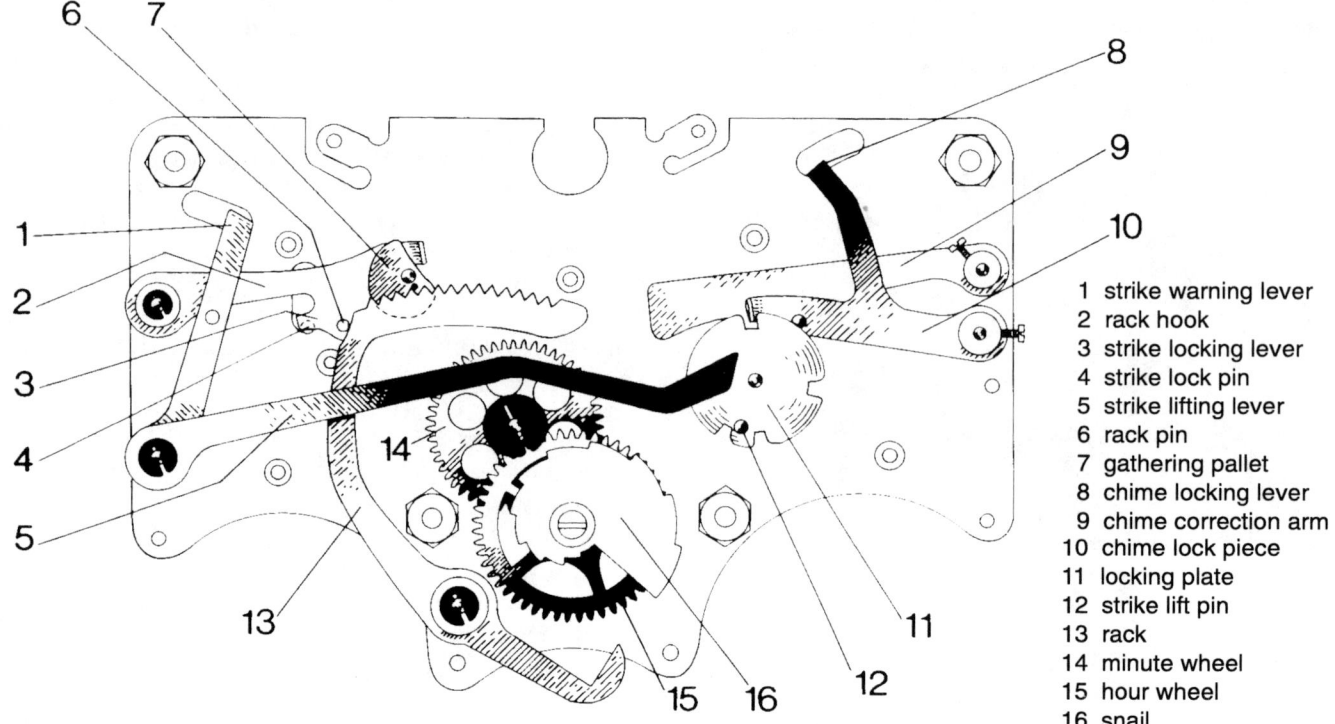

1 strike warning lever
2 rack hook
3 strike locking lever
4 strike lock pin
5 strike lifting lever
6 rack pin
7 gathering pallet
8 chime locking lever
9 chime correction arm
10 chime lock piece
11 locking plate
12 strike lift pin
13 rack
14 minute wheel
15 hour wheel
16 snail

Fig. 29. Seth Thomas No. 124 front movement.

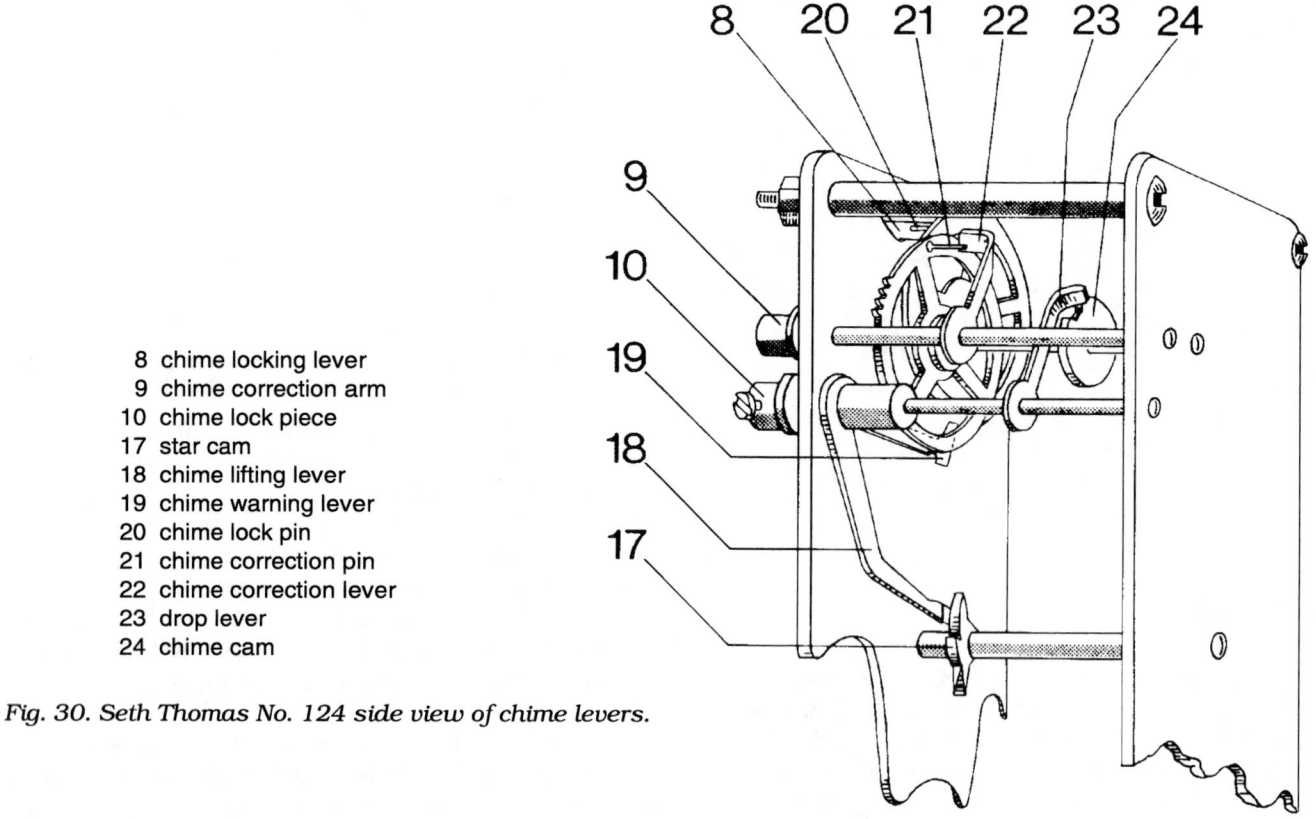

8 chime locking lever
9 chime correction arm
10 chime lock piece
17 star cam
18 chime lifting lever
19 chime warning lever
20 chime lock pin
21 chime correction pin
22 chime correction lever
23 drop lever
24 chime cam

Fig. 30. Seth Thomas No. 124 side view of chime levers.

CHAPTER 3 - SETH THOMAS NO. 124

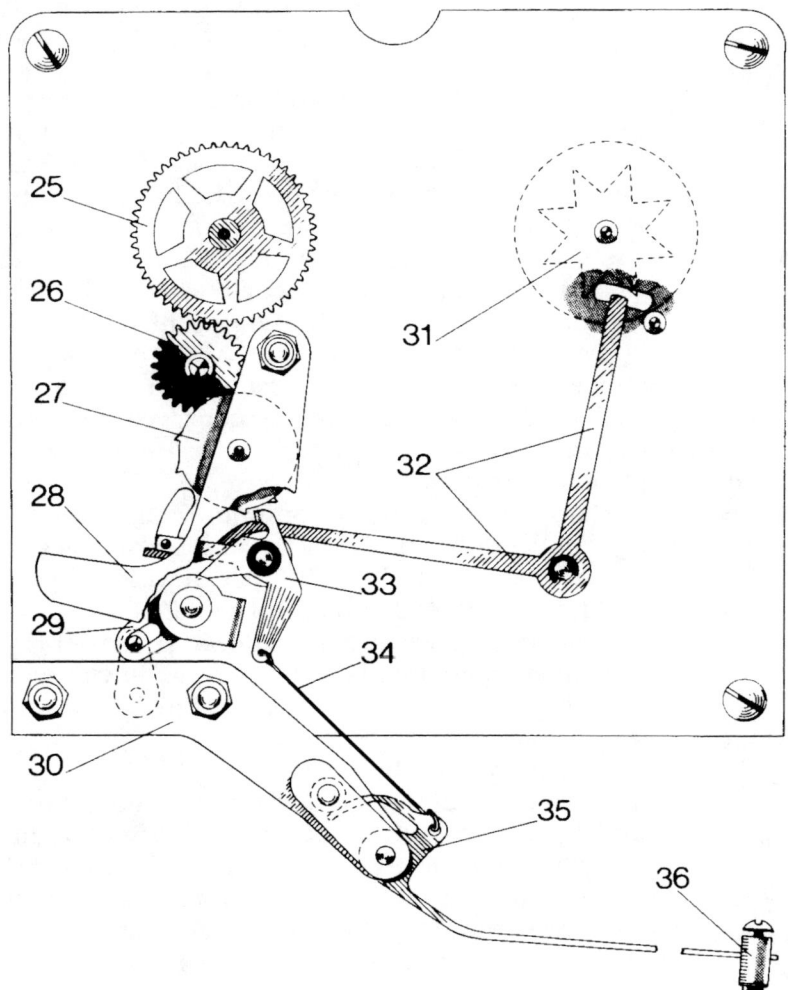

Fig. 31. Seth Thomas No. 124, rear movement.
- 25 chime drive wheel
- 26 idler wheel
- 27 pin barrel
- 28 chime bracket plate
- 29 chime silencer assembly
- 30 lower hammer bracket
- 31 hammer-lifting star and wheel
- 32 strike hammer-lift arm
- 33 hammer levers
- 34 hammer lift levers
- 35 lower hammer assembly
- 36 hammers

Fig. 32. Seth Thomas No. 124, alternate hammer arrangement for diagonal chime rods.
- 26 idler wheel
- 27 pin barrel
- 28 chime bracket plate
- 29 chime silencer assembly
- 32 strike hammer-lift arm (portion)
- 33 hammer levers
- 36 hammers
- 56 hammer spring

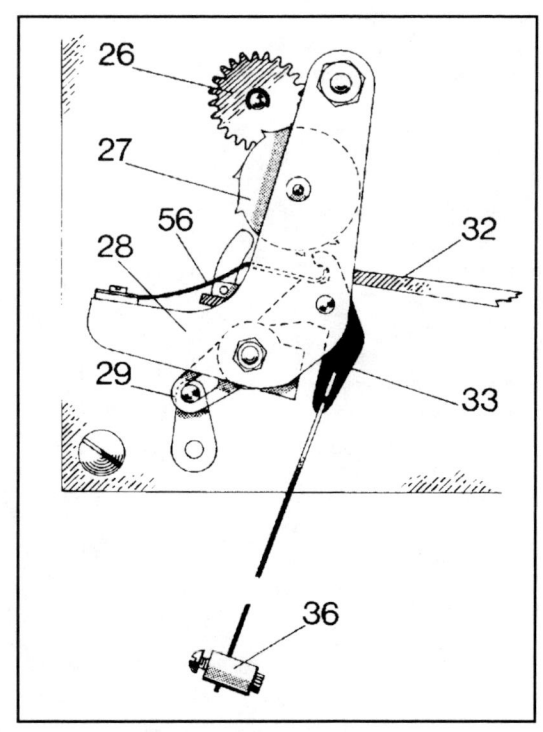

off the chime drive wheel (25), idler wheel (26), and the strike hammer-lift arm (32). Set aside the silencer lever spacer (42) and leather washer (43).

Now remove the upper front movement plate, leaving all the arbors standing in the rear plate. For identification, mark the strike second wheel, strike warning wheel, and strike fly. If you mark these parts, you should be able to identify all the others. Remove all the wheels carefully. Proceed with the cleaning of the movement parts. Do not neglect the mainsprings, which should be removed from the mainspring boxes. Always be careful when you handle springs. Thorough drying is important.

Next, check the pivots and holes. After the pivots have been polished, find out which holes require bushings. Fit all the arbors in place, a few at a time. It's a good way to become familiar with the movement; the assembly job later on will be easier. Install the bushings, and don't be surprised if quite a few are needed.

REASSEMBLY

Before you begin the reassembly, make sure that all six movement pillars are firmly screwed to the rear movement plate. The pillar near the center on the chime side is screwed to the pillar for the chime hammer bracket, with the rear movement plate held between them. Do not put in the silencer arbor or any other rear movement parts now, because they will interfere with the installation of the wheels. Place the rear movement plate on the bench, with the six movement pillars facing up.

You may want to separate the wheels into chime, strike, and time groups before you begin to install them in the rear plate. Based on the parts you have already marked, you should be able to identify all the wheels. Time train wheels, for example, do not have pins projecting from them. Escape wheel teeth have a saw-tooth appearance. The hammer-lifting star (31, Figure 31) identifies the strike third wheel. You will find the chime cam (24, Figure 30) on the same arbor with the chime fourth wheel. The chime third wheel has long front and rear pivots.

Do not forget to install the arbor carrying the chime correction lever (22), and the arbor with the chime lifting lever (18) and the drop lever (23). These two arbors cannot be added after the movement plates are together.

Carefully fit all the pivots into their holes, and tighten the pillar nuts only finger tight at this stage of the repair. Install all the rear movement parts, referring to Figure 31 for their locations. Do not add the mainspring boxes, main wheels, and lower front movement plate until later in the assembly process, after most of the chime and strike adjustments have been made.

On the front of the movement (Figure 29) install the rack hook (2), rack (13), cannon pinion, minute wheel (14), hour wheel (15), chime correction arm (9), chime lock piece (10), and strike lifting lever (5). Add the locking plate (11), but do not tighten the set screw. You are now ready to adjust the chime and strike mechanisms. It is easier to begin with the chime side.

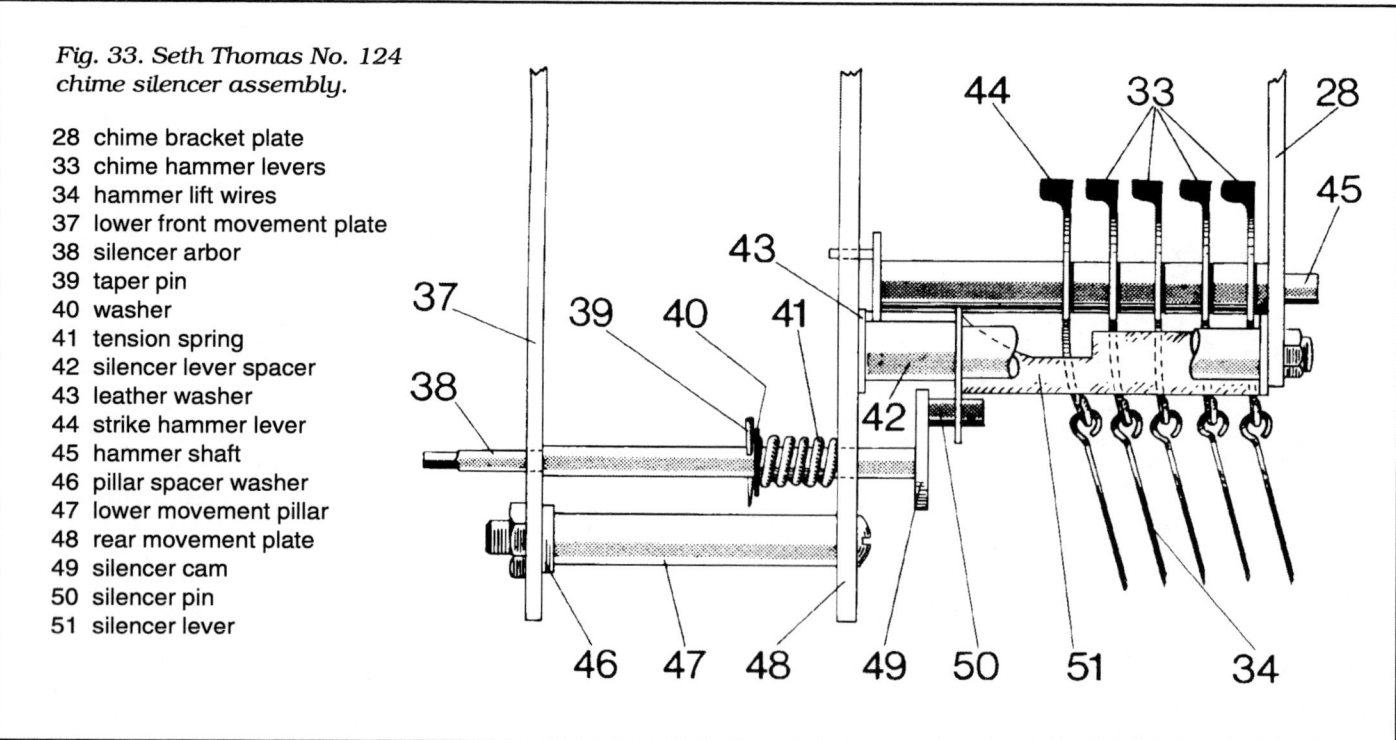

Fig. 33. Seth Thomas No. 124 chime silencer assembly.

- 28 chime bracket plate
- 33 chime hammer levers
- 34 hammer lift wires
- 37 lower front movement plate
- 38 silencer arbor
- 39 taper pin
- 40 washer
- 41 tension spring
- 42 silencer lever spacer
- 43 leather washer
- 44 strike hammer lever
- 45 hammer shaft
- 46 pillar spacer washer
- 47 lower movement pillar
- 48 rear movement plate
- 49 silencer cam
- 50 silencer pin
- 51 silencer lever

CHAPTER 3 - SETH THOMAS NO. 124

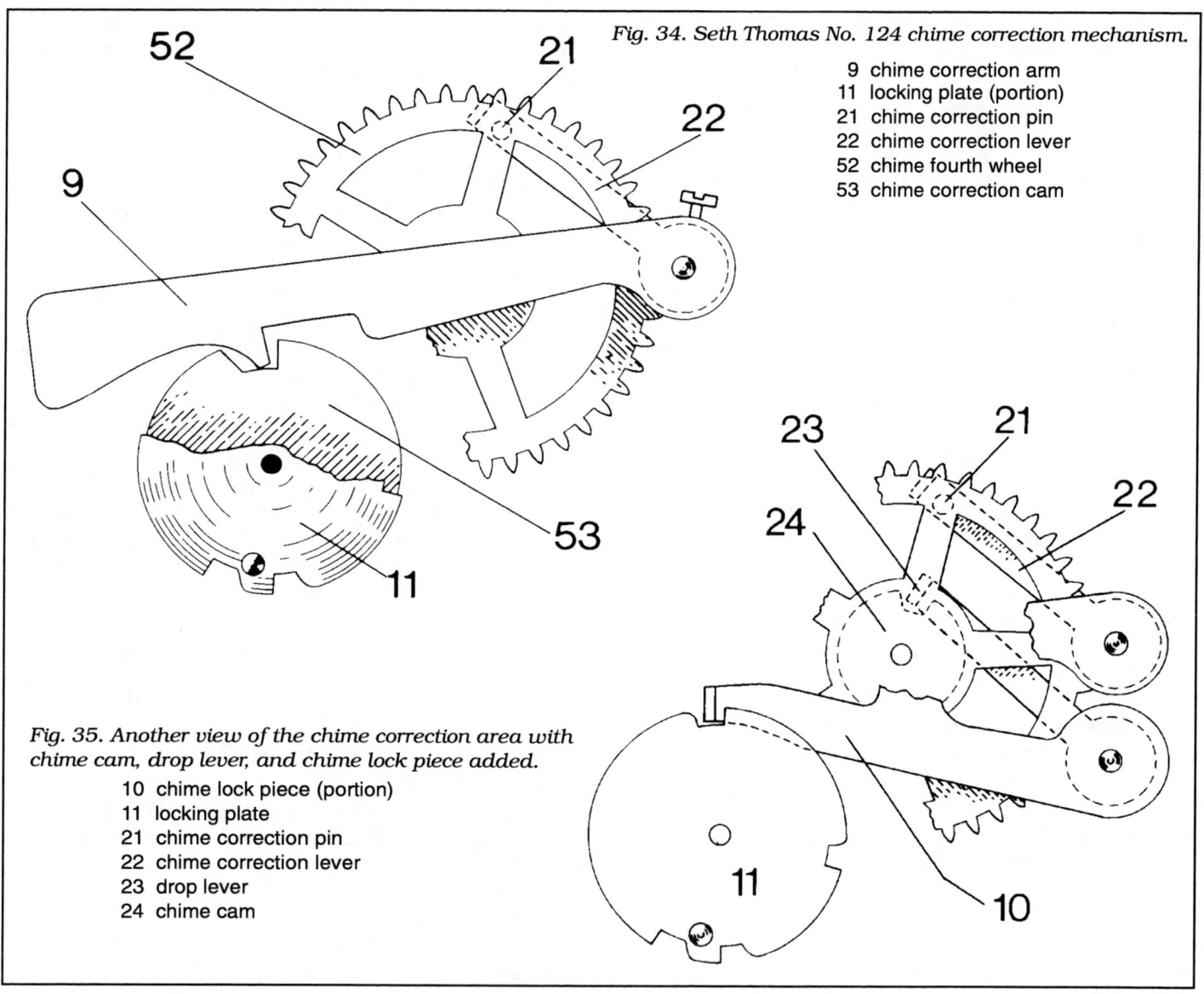

Fig. 34. Seth Thomas No. 124 chime correction mechanism.

9 chime correction arm
11 locking plate (portion)
21 chime correction pin
22 chime correction lever
52 chime fourth wheel
53 chime correction cam

Fig. 35. Another view of the chime correction area with chime cam, drop lever, and chime lock piece added.

10 chime lock piece (portion)
11 locking plate
21 chime correction pin
22 chime correction lever
23 drop lever
24 chime cam

CHIME ADJUSTMENTS

First, install the minute hand for testing purposes. Turn the chime wheels with finger pressure, with the fly rotating counterclockwise as seen from the front. When the drop lever (23) is centered over the slot in the chime cam (24), stop the gears. The chime lock pin (20) must be located at the chime locking lever (8) at this point. This synchronization is necessary for the chime train to run properly. Ease the plates apart enough to allow you to correct the mesh of the fourth wheel (52) with the pinion on the fifth arbor. Fit the plates together again and check the adjustment. Now turn the locking plate by itself until the chime lock piece (10) drops into the third quarter hour slot. Tighten the locking plate set screw. Double check the position of the chime lock pin, drop lever, and locking plate before going further.

Next, set the depth of the drop lever in relation to the chime lock piece. Unlock the gear train and turn it to mid-cycle. With the chime lock piece resting on the rim of the locking plate, place the drop lever on the rim of the chime cam. Tighten the set screw on the chime lock piece. The drop lever and chime cam can now perform their function. During the first few seconds of chiming, they keep the chime lock piece from falling back into the slot in the locking plate.

Proceed to adjust the chime correction mechanism shown in Figures 34 and 35. The chime correction arm (9) drops into the notch in the chime correction cam (53) at the end of the third quarter chime. The chime correction lever (22) must be set low enough to stop the chime correction pin (21). Adjust the lever to a trial position, and tighten the set screw on the chime correction arm. Turn the minute hand, and operate the gear train with finger pressure through several quarter hours. Observe the mechanism carefully, studying Figures 34 and 35 as you make adjustments.

Chime Correction Lever Too High

If the chime correction lever is set too high, it cannot function. No matter what happens, it will not stop the pin. Automatic chime correction is cancelled out. If the chimes get out of sequence, they stay that way. Any time the clock is allowed to run completely down, and any time the hands are turned either backward or rapidly forward, the chimes will be upset.

To correct the problem, loosen the set screw and move the lever (22) down. This changes its relationship with the chime correction arm (9). As shown in Figure 34, you can make the adjustment with the arm (9) in the slot. Note that with the parts arranged this way, the pin (21) has not rotated to a position to reach the lever (22). Just verify that the height of the lever is correct, so it will stop the pin during chime correction.

Chime Correction Lever Too Low

If the lever is too low, you may not get any chiming at all. Even at the hour, the extra lifting action is not enough to raise the lever (22) to release the pin (21). This defect is even worse than the one described above, because the chimes do not operate.

Correct the problem by taking the opposite action. Loosen the set screw, then raise the lever slightly. Run the chime train through several hours to check your adjustment. Upset the chime sequence, then check to make sure the chime correction mechanism will function. Then turn through another hour, to watch "normal" operation. At the hour, the mechanism should not interfere with the chime. If the lever is still slightly too low, it may not clear the pin on the first "pass." On the next hour, it may do so. Readjust if this is the case.

Install the lower front movement plate with spring boxes and main wheels attached. As you fasten the plate down, it will push on the tension spring (41) for the silencer arbor (38), creating the necessary stiffness in the chime silencer assembly. See Figure 33. Do not forget to install the pillar spacer washers (46), one on each of the two lower movement pillars. Tighten all six pillar nuts, and add the four plate screws which fasten the upper and lower front movement plates together. Wind the chime and strike springs partially.

Adjust the chime hammer sequences as explained in Chapter 1. The first quarter chime sequence, consisting of four descending notes, is played by the hammers operating in order from rear to front. Incidentally, the 1926 catalog offered the "Seth Thomas Chime" as an alternative to Westminster. This four-note sequence is similar to Westminster for adjustment purposes: the clock also plays four descending notes at the first quarter.

REPAIRING A DAMAGED NO. 124 CHIME CORRECTION DEVICE

After spending what seemed like a long time trying to adjust the chime mechanism of a Seth Thomas No. 124, I found that the clock still did not chime correctly. It was most puzzling. The clock chimed through an hour, but would fail to sound at the next hour. Then after remaining silent for four quarters, it chimed on the following hour. So it was: an hour "on", then an hour "off".

Figure 35 shows the parts involved. At first, I thought the problem might be in the drop lever (23) and chime cam (24). These parts are a source of confusion for many repairers, so it is appropriate for us to look at them. Their function is simple: during the first few seconds of chiming, they keep the chime lock piece (10) from falling back into the slot in the locking plate (11). The drop lever rides up on the cam, raising the chime lock piece. At the end of chiming, the drop lever must go into the slot in the chime cam. The drawing shows the parts in this position. It is important to note that the drop lever and chime cam do not lock the chime train; there is a separate chime locking lever.

If I had noticed the drop of epoxy on the 4th arbor earlier, I would have had the answer to my problem. The chime cam (24) had been moved out of position and then cemented in place. When chiming ended, the chime correction pin (21) stopped about 30° away from the chime correction lever (22). This was the key to the "off and on" chiming.

At the hour, the chime correction mechanism lifted the lever (22) out of the way of the chime correction pin (21) and chiming began. Before the first note, however, the pin and lever came together, stopping the chime train. On the following three quarters, the clock was silent, because it was waiting to chime the hour with the minute hand pointing up. Then, at the hour, it released for the chime. But when chiming stopped, the pin ended up 30° away from the lever again. So we were silent again for an hour.

The chime correction pin should be located close to the lever. At the hour, when the lever is raised, the pin slides under and past it as chiming starts. To assure a correct position, the slot in the chime cam (24) must point directly at the pin (21) as shown in Figure 35. The two rotate together, one on the arbor and the other on the wheel.

This "special case" helps to illustrate the way the chime correction mechanism works. The chime

CHAPTER 3 - SETH THOMAS NO. 124

correction pin stops close to the chime correction lever, ready to halt the chime train if necessary. In addition, the example helps to clarify the role of the chime cam as a "start-up" device for the chime train. The chime cam played an important part in the example because it was out of position. The whole chime train was adjusted in relation to it, and this put the chime correction pin out of place.

"RARE" CHIME CORRECTION DEVICE

There is a Seth Thomas No. 124 movement with a variation of the chime correction mechanism. The "variation" seems to represent a relatively small number of movements. Upon checking Seth Thomas catalogs dated 1924 to 1928, I found pictures of the movement with the variation rather than the common chime correction mechanism. This may mean that the early design was dropped after only a few years of production. Besides the catalog illustrations, two other thoughts seem to bear this out. First, the "variation" resembles the setup in the older No. 113 chime; and second, the mechanism just isn't as good as the more common one.

The Two Chime Correction Devices

Figure 36 shows the two devices. Views A and B show the same upper right corner of the movement. Variation "A" is the mechanism pictured in the 1924-28 catalogs. "B" is the common chime correction mechanism described earlier in this chapter. I'll compare them briefly before leaving "B" aside.

"B" is the common chime correction device we know from so many No. 124 repairs. The chime correction arm (in black) rides on a cam hidden behind the circular locking plate. Following the third quarter chime, the arm drops. It allows the chime correction lever to stop the chime correction pin. At the hour, the lifting action is high enough to lift the lever out of the way. A set screw holds the chime correction arm and lever at exactly the right angle. There is very little leeway permitted in making the adjustment. The automatic chime correction may not work, or the clock may not chime at all. Repairers often cancel out the device to get the chimes working again.

In "A" the mechanism is different. Note the two levers drawn in black. The one on the right is the chime correction arm, and the lever on the left holds it in place to correct the chimes. There is no cam behind the locking plate, and no set screw for adjustment. This device is even more of a weak link than the common mechanism shown in view "B." Figure 37 illustrates the main parts of the "A" version. The back surface of the locking plate (11) carries a pin. Pin and locking plate rotate clockwise as the movement chimes. As the third quarter chime ends, the pin pushes down on the chime correction arm (9). This action places the chime correction lever (22) in the path of the chime correction pin (21). If the chimes are out of sequence, the lever and pin will stop the chime train from any further operation.

The retaining lever (58, in black) holds the chime correction arm in place during the correction phase. It pivots at the left, and is moved by gravity only: no spring. The pointed tip drops into a notch in the chime correction arm as the arm moves down. Chiming cannot occur as long as the retaining lever locks the arm in place. At the hour, the lift is high enough to raise the chime lock piece (10) and release the chimes. A pin (shown in black on the chime lock piece) pushes up the retaining lever (58), mov-

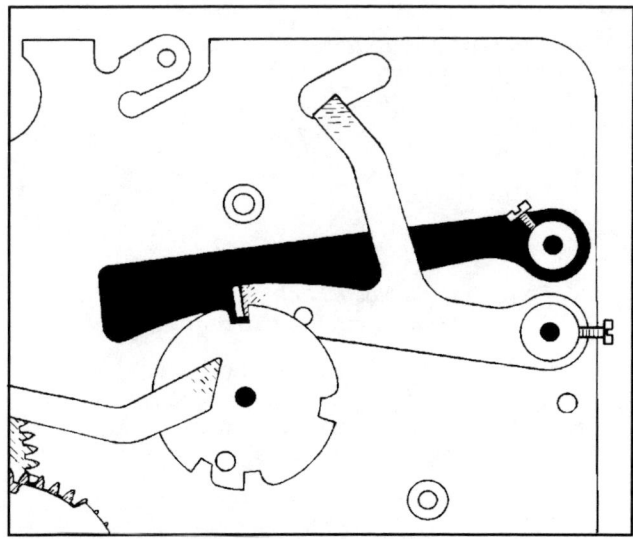

A The variation.   B Common version used in most No. 124 movements.

Fig. 36. Seth Thomas No. 124 chime correction devices.

ing it out of the notch in the chime correction arm (9). Now the chime correction spring (57) moves the chime correction arm and chime correction lever up together. The spring is all-important to the mechanism, for without it the device cannot work. Wrapped around the common arbor for parts 9 and 22, the spring is anchored around the movement pillar. As you view the clock from the front, the spring action pivots the parts clockwise.

The inset view in Figure 37 shows the end result. The tip of the retaining lever (58) has popped up out of the notch. Under firm spring pressure, the chime correction lever (22) has moved upward. The chime correction pin, shown at the arrow, is now free to move past the lever as the hour chime begins. The levers will stay this way until the end of the third quarter chime arrives again. Then we're back to Figure 37, main view, with the device locked in a correction phase. If the chimes are still properly synchronized, the chime correction lever will never touch the pin. The lever only stops it if correction is necessary.

Adjustment and Repair of the "A" Device

To check the operation of the "A" chime correction device, first look at the spring (57). It must be in good condition, without kinks or bends. It makes sense to replace it so it will not fail. Press down on the chime correction arm (9), and it should snap back up under spring pressure. Check the direction of the pressure, because it can easily end up being the wrong way if the spring is old or distorted. Another way to look at it is to realize that the spring holds the device in the "run" position. The retaining lever only holds it temporarily in the "correction" position.

Next, watch the retaining lever (58) as it drops into the slot in the chime correction arm (9). A worn or damaged notch will not hold the arm against the pull of the spring. A movement I studied had this problem. The notch was worn to the point that the retaining lever could only hold the arm on about half the tries I made while testing it.

To repair the notch, remove the chime correction arm (9) from the movement. It is not possible to get a file on the arm, even if the movement is apart, unless you take the arm off the arbor. I found the arm to be a press-fit, not difficult to remove. Hold the arm in a vise and carefully clean up the notch with a file to make the edges sharp and smooth. Try not to change the angle. If the notch is opened up too much, it will not be able to hold the lever; if too deep and wedge-shaped, the notch will jam the retaining lever (58) as soon as it is seated.

A few other things should be checked. The retaining lever must pivot freely on the post. The pointed end should be clean and reasonably sharp.

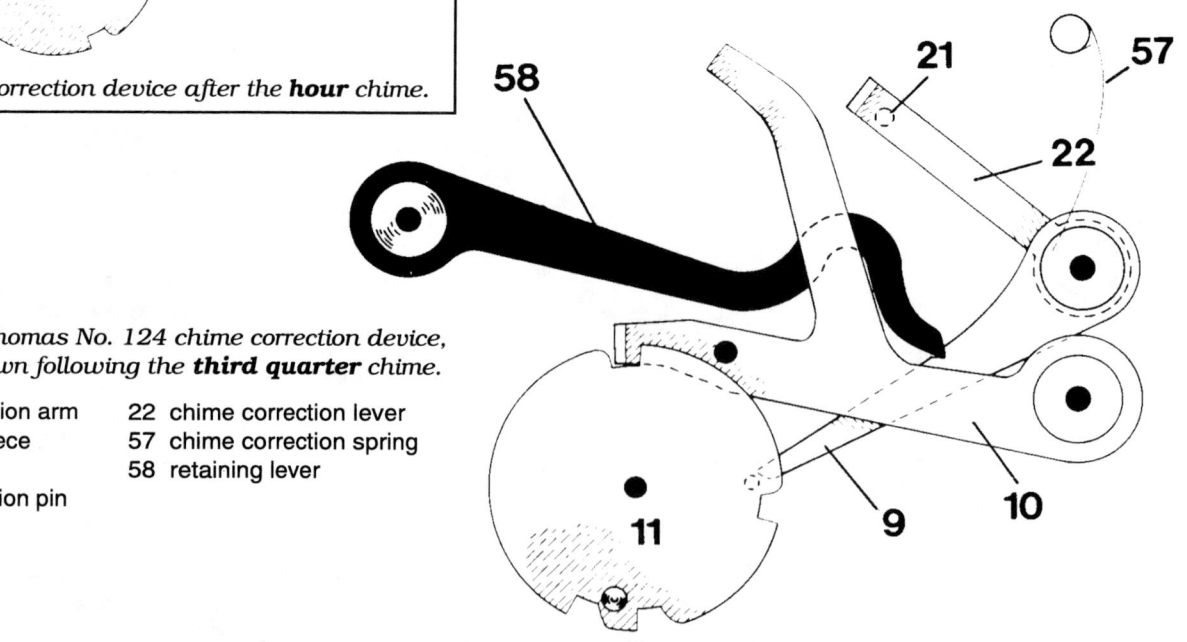

Fig. 37. Seth Thomas No. 124 chime correction device, version A, shown following the **third quarter** chime.

- 9 chime correction arm
- 10 chime lock piece
- 11 locking plate
- 21 chime correction pin
- 22 chime correction lever
- 57 chime correction spring
- 58 retaining lever

*The chime correction device after the **hour** chime.*

CHAPTER 3 - SETH THOMAS NO. 124

Watch as the pin on the locking plate comes around to move the arm. It should contact the arm and push it down far enough to allow the retaining lever to drop all the way into the notch.

A check should be made on the height of the chime correction lever (22). Assuming the chime correction arm is properly installed on the arbor, the height for the lever should be correct. If it is not, move the arm on its arbor, changing the angle between parts 9 and 22. Observe the chime correction mechanism to see if it works after the adjustment. The details for the adjustment are covered earlier and illustrated in Figures 34 and 35.

STRIKE ADJUSTMENTS

Proceed to adjust the strike train, referring to Figures 29, 31, and 38 as you work. Check to make sure that the strike hammer-lift arm (32) is not in contact with the hammer-lifting star (31). If there is contact when the hammer should be at rest, the hammer will always finish the strike cycle in a partially raised position. The strike train may stall as a result of the extra starting resistance. The gear train must be able to reach full speed before the hammer-lifting star engages the weight of the strike hammer. If you need to correct the position of the hammer-lifting star, carefully rotate the gathering pallet on its pivot (see Figure 38). Take care to verify that the gathering pallet is loose enough to turn, so you will not bend or break the pivot. Try different locations until the strike train locks with the hammer lifting star clear of the strike hammer-lift arm. Make sure that the clearance is maintained even after the strike warning occurs.

Next, check the strike locking. Start the gear train, but stop the fly (55) at the instant the rack pin (6) drops off the end of the rack, near the completion of the cycle. Note the exact position of the strike lock pin (4) at this point. Then allow the gears to turn very slowly around to the lock position. The strike lock pin will make less than one revolution before hitting the strike locking lever (3). If the pin moves at least one third of a revolution, the locking

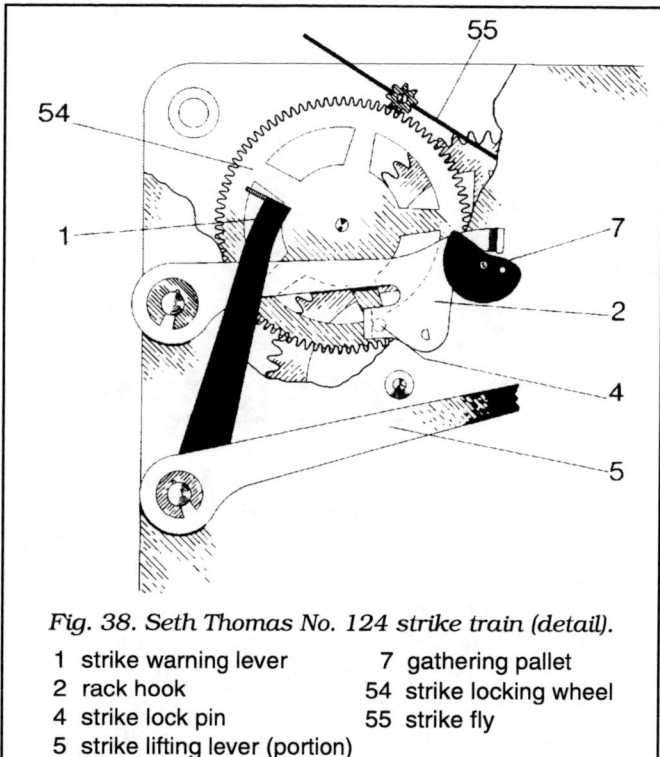

Fig. 38. Seth Thomas No. 124 strike train (detail).

1 strike warning lever
2 rack hook
4 strike lock pin
5 strike lifting lever (portion)
7 gathering pallet
54 strike locking wheel
55 strike fly

is safe. However, if there is little or no "run" of the pin, a problem can develop. The lock pin may glance off the tip of the descending strike locking lever, causing mislocking. An extra hammer blow will be struck before the train locks on the next pass. To correct the adjustment, rotate the gathering pallet on the pivot again, but very slightly. Repeat until the locking action is safe. Complete the strike train adjustments by checking the operation of the rack and snail at 12 o'clock and 1 o'clock.

After the chime and strike trains have been adjusted and all other required work is done, the Seth Thomas No. 124 movement is ready for testing. Time train function must be checked, of course, for proper escapement action, regulation, and eight-day running time. The chime and strike trains must also work for the eight days, without errors in sequencing or counting.

4

SETH THOMAS NO. 113

If we are to view the No. 113 chime movement in proper perspective, it should be compared to the No. 124 from Chapter 3. The No. 113 is a larger movement, which Seth Thomas described as the "high grade clock movement" in its catalogs of the 1920's. The No. 124 is a less expensive movement characterized as "dependable," a "smaller, more compact movement built to do the same work as the No. 113". Throughout the 1920's and '30's, these two movements formed the basis for the Seth Thomas keywound chime clock line. The No. 113 appeared about 1921; the No. 124 arrived in 1924. It appears the latter was made in far greater quantities for lower-priced clocks.

The No. 113 was produced in several pendulum lengths. It was fitted in large tambours such as the No. 75 from the 1924 catalog. The description says the 22-1/2 inch-wide clock is "well adapted to club rooms and pretentious private interiors." There were also the large mahogany "cabinet cases." Figure 39 shows two of them, Numbers 72 and 73. These had regulator and chime silencer levers placed in the upper section of the dial. Wall and floor clocks were also offered with the No. 113 chime movement. The Hall Clock No. 3W from 1939, priced at $125.00, was 73 inches tall.

Figure 40 shows two versions of the No. 113 movement. The No. 113A (large view) is from the 1926 Seth Thomas catalog. Note the double clicks placed at each winding position. These help to prevent accidental release of the mainsprings from a click spring failure, reducing the risk of serious movement damage. The 1921 catalog picture (not shown) has single clicks, so it appears the double clicks were added soon after the No. 113 was introduced.

DISASSEMBLY AND REPAIR

Restoration begins with the letting down of the three mainsprings. Remove the geared winding mechanisms, barrels, and lower front movement plate.

Fig. 39. Seth Thomas Chime Clock No. 72 (left) and No. 73 (right), each with No. 113 chime movement.

Each gear train has its own barrel and cover, mainspring, barrel arbor, winding arbor, 16-tooth winding wheel, 26-tooth click wheel, and cover plate to hold the geared winding mechanism in place. The gearing provides a ratio of better than 1.6:1, a useful advantage in winding the powerful mainsprings. Mark the barrels, covers, and springs so you can identify them later. The time and strike barrel arbors are the same; the odd one belongs to the chime. All the other winding parts listed above are interchangeable.

After the winding mechanisms and barrels are removed and put aside for cleaning, look the movement over for worn pivot holes. Apply pressure to the wheels to cause the pivots to move from side to side. Note the locations of worn holes to be bushed after the movement has been taken apart and cleaned. Bear in mind that heavy dirt can conceal worn holes. You must check the movement after cleaning. Polish all pivots on the lathe.

It is a good idea to mark the three second wheels as chime, strike or time before you disassemble. The other wheels are easily identified. The chime fourth wheel, for example, is on the same arbor with

CHAPTER 4 - SETH THOMAS NO. 113

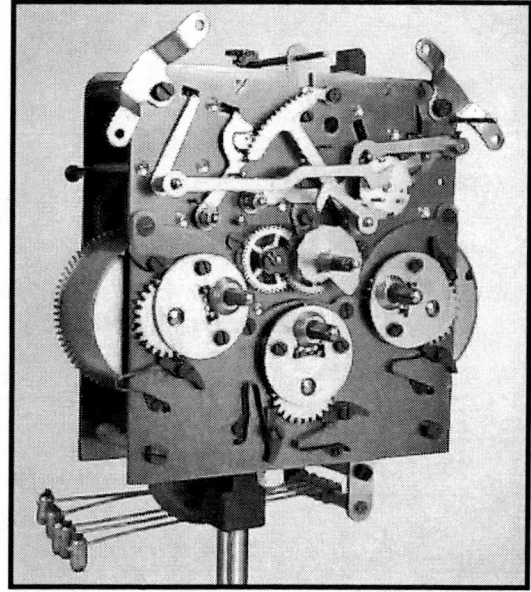

Seth Thomas No. 113 chime movement from Chime No. 78 tambour clock. This No. 113 is designed to play the "Seth Thomas Chime" instead of Westminster.

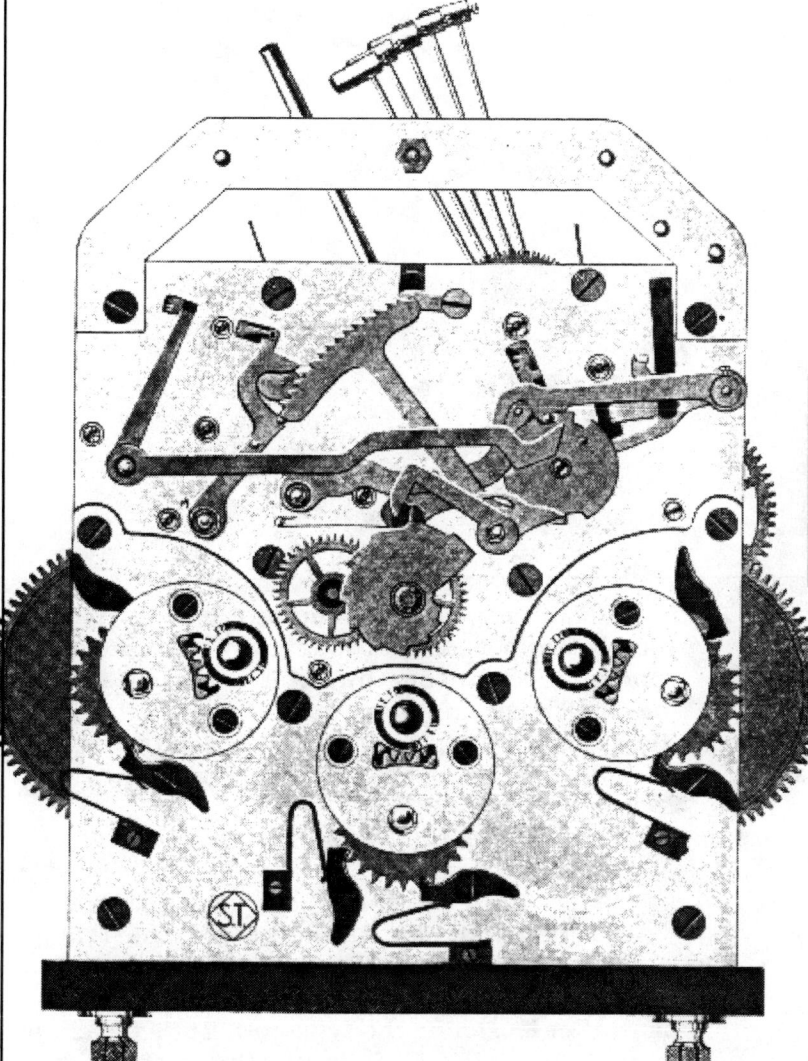

Mainsprings for No. 113 Movement
Time and strike: 7/8" wide x .018" thick,
  calculated length: 76"
Chime: 1" wide x .018" thick,
  calculated length 95" (can use 96")

Fig. 40. Two versions of the No. 113 chime. Left: Westminster chime movement No. 113A from the 1926 Seth Thomas catalog, illustration courtesy Tran Duy Ly, Arlington Book Co. Top of page: No. 113, author's photo.

A HIGH-GRADE Westminster chime movement consisting of three separate trains to operate the hour strike, the time and the chime. All wheels are accurately cut of heavy brass, solid cut steel pinions, burnished and polished; pivots hardened and burnished. Heavy cut steel pallets perfectly ground, hardened and polished to reduce friction and insure long life.

The plates made of heavy hard rolled brass. All pivot holes provided with deep cut oil cups for retaining oil. Fly staff pivots and pallet staff are adjustable by moving adjusting buttons set in front plate. Large mainspring barrels which can be easily removed by detaching lower section of front plate without taking movement apart.

Each winding ratchet fitted with double clicks.

A reliable chime adjustment or self-correcting lever which regulates or corrects the chime and hour strike automatically within the hour. The chimes may be silenced by moving the indicator hand, allowing the hours only to strike.

The movement is rigidly mounted on a seatboard securely fastened to clock case, and may be removed by detaching hands and loosening two thumb screws.

A movement of the finest design and workmanship acquired by our experience for over 100 years in clock manufacturing.

In Chime Nos. 72, 73, 74, 75 and 78.

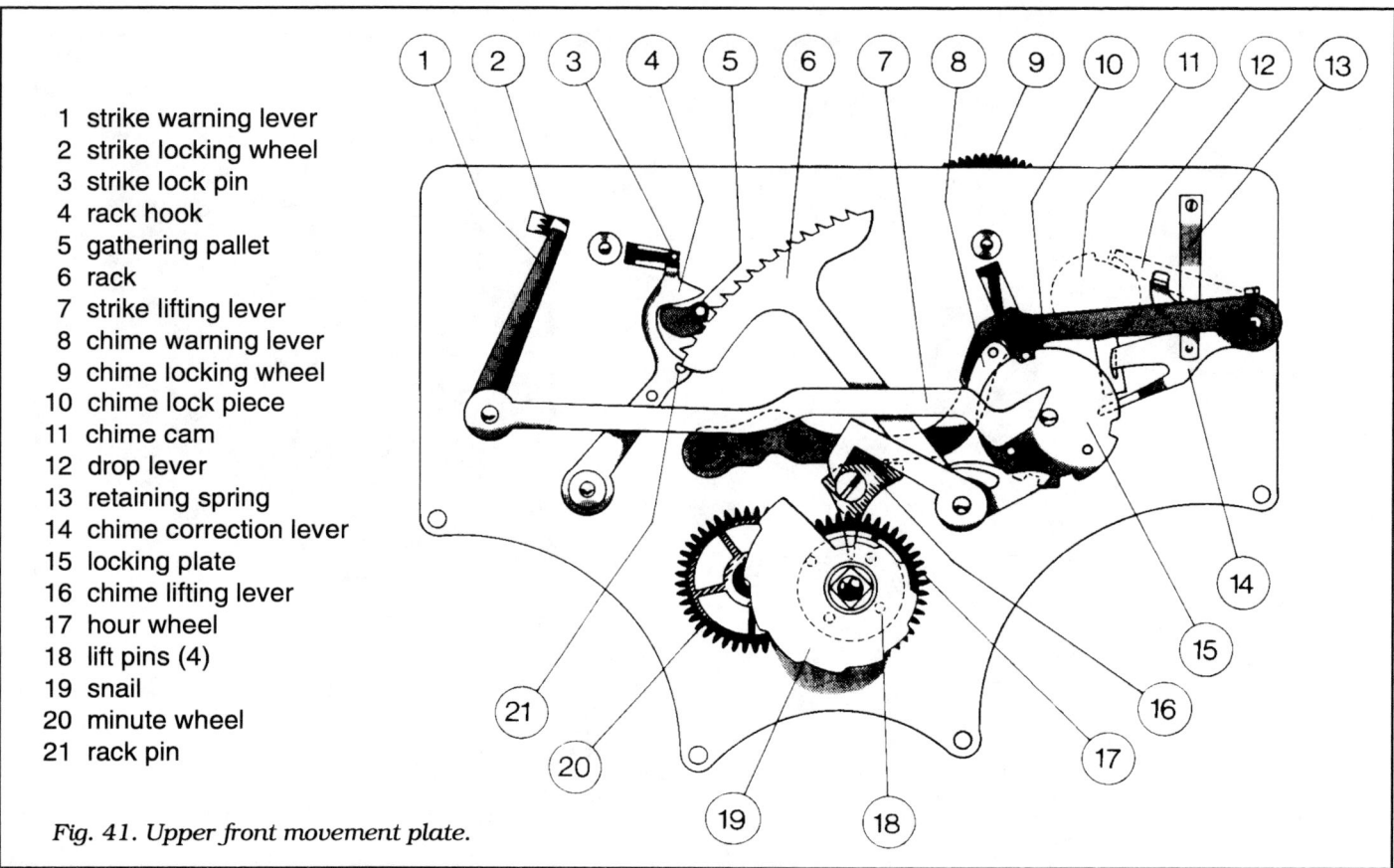

1. strike warning lever
2. strike locking wheel
3. strike lock pin
4. rack hook
5. gathering pallet
6. rack
7. strike lifting lever
8. chime warning lever
9. chime locking wheel
10. chime lock piece
11. chime cam
12. drop lever
13. retaining spring
14. chime correction lever
15. locking plate
16. chime lifting lever
17. hour wheel
18. lift pins (4)
19. snail
20. minute wheel
21. rack pin

Fig. 41. Upper front movement plate.

the chime cam, and the strike third wheel shares the arbor with the hammer-lifting star. Many repairers do not disassemble chime clocks because they are afraid they will mix up the parts and never be able to sort them out again. But it is relatively easy to know all the wheels if a few are marked and the rest identified by special features.

CHIME

Reassembly and adjustment are the real challenges of the repair. Refer to Figures 41 and 42 to locate the various front movement parts. The movement bears some resemblance to the No. 124 covered in Chapter 3; the similarity is easier to see without the lower front movement plate barrels and winding mechanisms in place. As you install the wheels and arbors into the back plate, remember to add the arbor which carries the chime locking lever (23). The hammer arbor must also be added at this stage.

The silencer arbor can be saved for the end of the assembly procedure, after the plates are together. In tambour models like the Chime No. 74 and No. 78, the silencer mechanism works much like the No. 124. No. 113 movements in the larger "cabinet cases" feature the horizontal bracket shown in Figure 40 on top of the No. 113A movement. This carries the silencer and regulator arbors.

Before tightening down on the upper front movement plate, check the chime train for correct assembly. Do not install the mainspring barrels and winding mechanisms until you are finished working on all basic adjustments. If the chime train is in the locked position, three conditions must be met:
1. The chime lock pin (22) is stopped at the locking lever (23).
2. The drop lever (12) is in the slot of the chime cam (11).
3. The pin in the chime lock piece (10) is in one of the four slots in the locking plate (15).

It is easy to adjust the position of the locking plate after loosening the set screw. But if the chime lock pin is not synchronized with the chime cam, you must partially separate the movement plates and change the mesh of the wheel and pinion connecting these two parts.

Adjust the angle between the chime lock piece (10) and the drop lever (12). Start the chime cycle under finger pressure. When the drop lever is riding on the chime cam (11) and not in the slot, tighten the set screw on the chime lock piece. Check that the chime lock piece rides on the locking plate, and the drop lever rides on the chime cam. This adjustment is necessary so locking will occur at the end of the cycle.

It is important to understand the chime locking

CHAPTER 4 - SETH THOMAS NO. 113

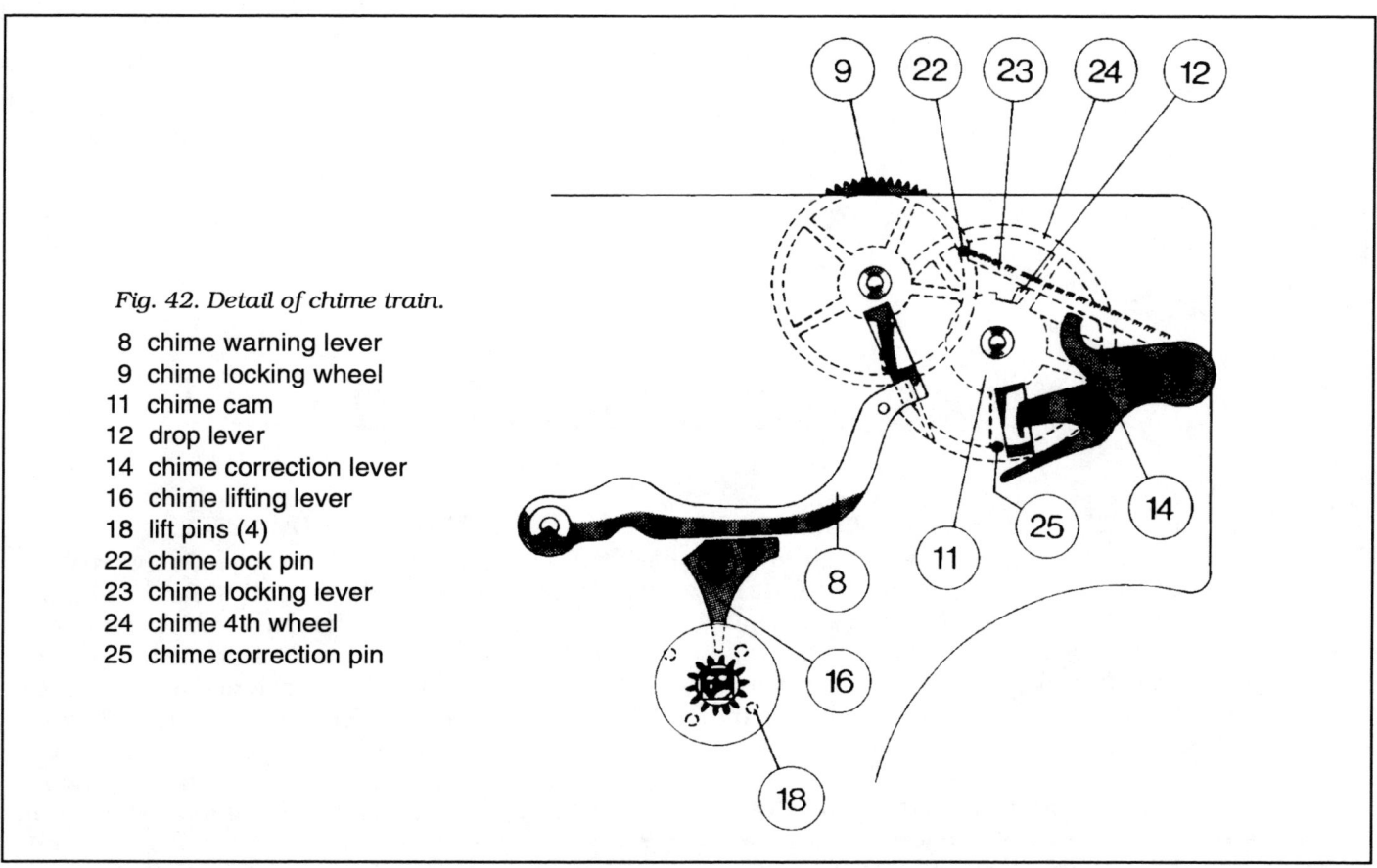

Fig. 42. Detail of chime train.

- 8 chime warning lever
- 9 chime locking wheel
- 11 chime cam
- 12 drop lever
- 14 chime correction lever
- 16 chime lifting lever
- 18 lift pins (4)
- 22 chime lock pin
- 23 chime locking lever
- 24 chime 4th wheel
- 25 chime correction pin

function. The locking is accomplished by the chime lock pin (22) and the chime locking lever (23). The drop lever doesn't have a locking function, although locking cannot happen unless it is in the cam slot. The drop lever and chime cam keep the chime lock piece out of the slot in the locking plate as chiming begins. Similarly, the locking plate itself is not for locking. It is a counting device which locates the end of each Westminster chime note sequence.

Now look at the automatic chime correction feature. It restores the chimes to the correct sequences by means of the chime correction lever (14). The retaining spring (13) holds the lever in the raised or lowered position. As the third quarter chime ends, a pin on the locking plate pushes the chime correction lever to the "down" position. It is held there by the spring. The chime correction lever is now in the path to be followed by the chime correction pin (25) on the chime 4th wheel (24). If the next chime note sequence is not the hour, with the minute hand pointing straight up, the chime correction pin will prevent the gear train from running. At the actual hour, the lift pin for the hour is the only pin out of the four pins (which are shown as #18) able to lift the levers high enough to move the chime correction lever up again. It is now out of the way of the chime correction pin, and the hour chime begins.

Under synchronized chiming, the chime correction lever is pushed down as the third quarter chime ends and back up 15 minutes later, without ever interfering with the chimes.

STRIKE

The strike lock pin (3) is held by the rack hook (4) to lock the gear train. At the warning, near the end of the hour chime, a pin on the locking plate raises the strike lifting lever (7), releasing the locking wheel (2) and strike lock pin to run to the strike warning lever (1).

Although the strike lock pin does both the locking and warning, it still must be set. Loosen the gathering pallet (5) and move it so the hammer tail is clear of the hammer-lifting star when the train locks. Closely observe how the strike lock pin operates. When the rack hook rests in the notch in the gathering pallet, locking should occur. Further, the rack hook should be fully in position before the pin hits it to lock the train. You can adjust by separating the plates enough to permit changing the position of the locking wheel. As an alternate, move the gathering pallet an additional small amount instead. Be careful with the gathering pallet: it must be loose enough so you will not snap the pivot. Push it on firmly after all the adjustments are completed.

5

SETH THOMAS SONORA CHIME

Most collectors and repairers know of the Seth Thomas Sonora Chime. Each clock has two movements in the case, one for the time and hour strike and the other for the quarter hour chimes. It is a curious design, and it means more work for the repairer. In a successful repair, the two movements must first be placed in good working order. Then they must be coordinated and adjusted to work together.

The Sonora model was introduced about 1909. The factory made a number of versions during the second decade of the 20th century, all with the same basic movements. Some chimed on bells, others on rods; there were Westminster and dual chime models. Many different cases featured the Sonora Chime, including round top, Gothic, tambour, and the black mantel type. In the 1920's, the three train No. 113 and No. 124 chime movements replaced the Sonora line. (See Chapters 3 and 4 for repair procedures on these two movements.)

The time and strike movement in the Sonora Chime clock is the No. 89 movement. It is a common count wheel style clock movement found in many other Seth Thomas clocks of the period. The factory modified it for the Sonora application. Every quarter hour, the No. 89 releases the separate chime movement to run. As the hour chimes finish playing, the chime movement then "returns the favor" by unlocking the No. 89 to strike the hour.

Within the clock case, the chime movement is located behind and toward the time side of the No. 89. It is powered by a heavy mainspring in a barrel. Chime winding is at the dial near the 3 o'clock position. Most Sonora models have a silencer lever at the side of the case. This lever raises the hammers, preventing them from reaching the bells or rods. Counting of the chime notes is controlled by a locking plate. There is no automatic chime correction feature. If the chime sequence is upset for any reason, it will not correct itself.

SONORA CHIME MOVEMENT

Several versions of the chime movement were made for rod chimes, bell chimes, and different hammer arrangements. The mechanism itself remained essentially the same. Figures 43 and 44 show the No. 119 chime movement as designed to chime the Westminster melody on four rods. This model strikes the hour on three of the same rods.

In Figure 43, the chime movement is shown in the locked position as it appears after the third quarter. It is important to note that locking is accomplished by the chime locking lever (5), held in the slot of the chime cam (3). There is no locking pin on any of the wheels.

The chime cycle begins a few minutes before each quarter hour. The chime lifting lever (27), shown in Figure 45, is part of the No. 89 time and strike movement. The lever contacts the chime unlocking lever (9, Figure 43) and lifts it. In turn, the chime warning lever (8) raises the chime locking lever out of the slot in the chime cam, releasing the gear train for the warning. Warning is achieved as the chime warning pin (6), mounted on the fly arbor, hits the chime warning lever to stop the gear train. At this point the fly has turned only about 1/2 to 3/4 revolution. The chime lock piece (4) has come up out of the slot in the locking plate (2). In addition, the chime cam has turned counterclockwise a few degrees.

Exactly at the quarter hour, the chime lifting lever on the No. 89 snaps back down to its pre-warning position. The warning lever then drops downward, releasing the warning pin and the entire gear train. Chiming begins. As the chime cam turns, the chime locking lever remains just above the edge of the cam. It does not fall into the cam slot during chiming. If it did, the gear train would be stopped in the midst of a note sequence. It is the locking plate which prevents this from happening. The chime lock piece rides along on the rim of the rotating locking plate, and it is held high enough so the

CHAPTER 5 - SETH THOMAS SONORA CHIME

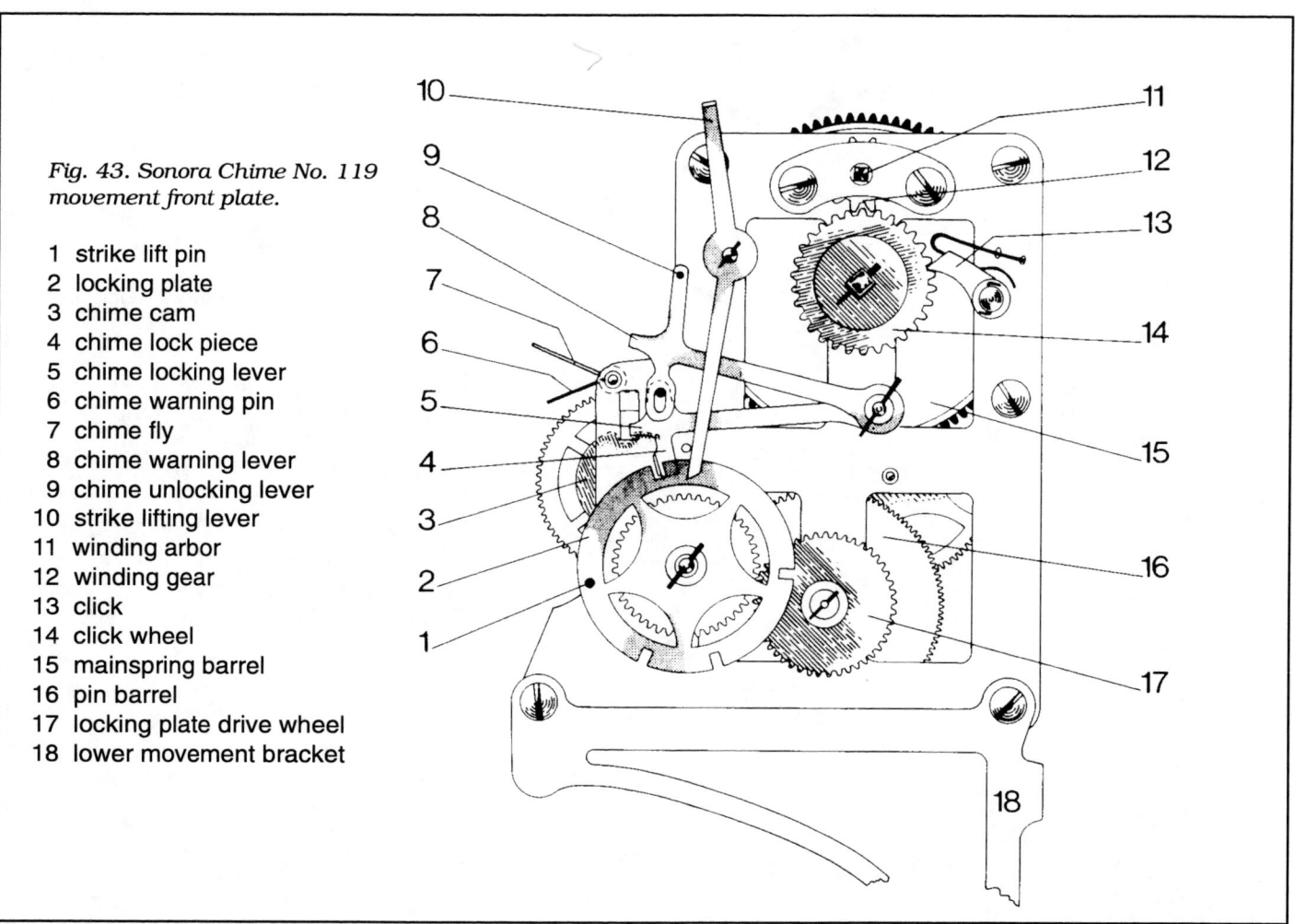

Fig. 43. Sonora Chime No. 119 movement front plate.

1 strike lift pin
2 locking plate
3 chime cam
4 chime lock piece
5 chime locking lever
6 chime warning pin
7 chime fly
8 chime warning lever
9 chime unlocking lever
10 strike lifting lever
11 winding arbor
12 winding gear
13 click
14 click wheel
15 mainspring barrel
16 pin barrel
17 locking plate drive wheel
18 lower movement bracket

chime locking lever is kept out of the slot in the chime cam.

The chime sequence ends as one of the four slots in the locking plate comes into position under the chime lock piece. Rotating in synchronization, the chime cam appears with its slot under the chime locking lever at the same instant. Both levers fall into their respective slots, locking the chime train. Chime note counting is controlled by the spacing of the slots in the locking plate. There are four progressively longer intervals corresponding to the 4, 8, 12, and 16 notes played at the quarter hours. Pins on the pin barrel (16) lift the hammers.

The easiest way to adjust the chime movement is to put it together correctly in the first place. That's another way of saying that some adjustments are difficult to make after the movement is completely assembled. If you assemble the movement in random fashion, you may end up tearing it down again to correct an error. With this in mind, we will go through assembly and adjustment as one procedure instead of two separate steps. Refer to Figures 43 and 44 showing the front and rear of the chime movement.

Let's assume the movement is apart, completely cleaned including the mainspring, with bushings installed as necessary. The pivots are in good condition as well. Now the movement is ready to be reassembled. First, screw the movement pillars into the rear plate. Install the second wheel, then the barrel (15), followed by the rest of the wheels, into the rear plate. The complete hammer assembly (parts 22 through 26) and the lower movement bracket (18) must go in now. Make sure you add all the original spacers and washers within the hammer assembly on your movement. The number and sequence of these washers varies from one model to another.

Before you install the front plate, put the chime lock piece and chime locking lever (4 and 5) and the chime warning lever (8) on it. Having these parts in place now will help you to begin adjusting the movement even as you fit the plates together. Begin to install the front plate now. As you work at getting the pivots into the holes, bring the chime cam (3) around so the chime locking lever goes into the slot. This puts the gear train in the locked position. Set the fly pinion in mesh with the chime locking wheel

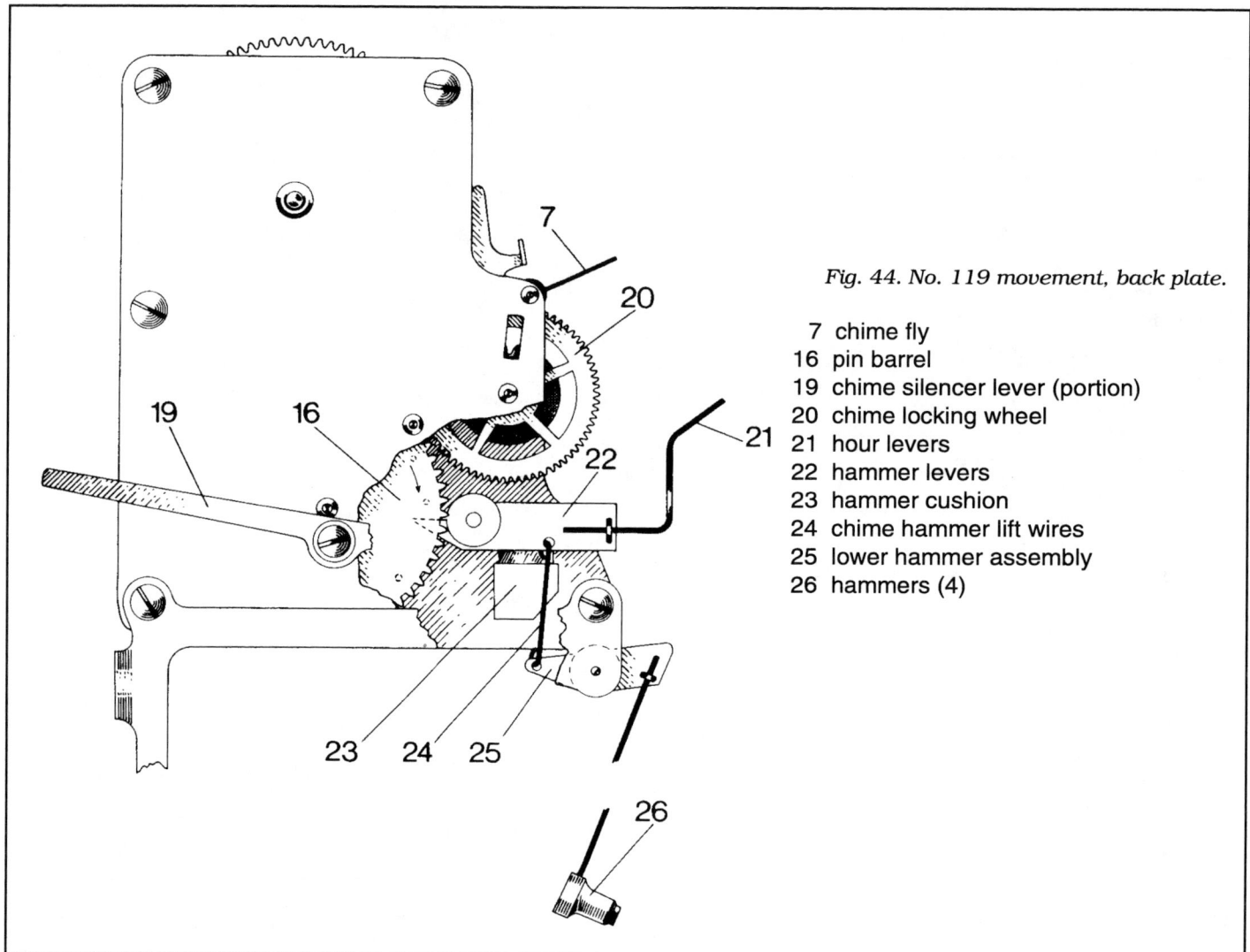

Fig. 44. No. 119 movement, back plate.

- 7 chime fly
- 16 pin barrel
- 19 chime silencer lever (portion)
- 20 chime locking wheel
- 21 hour levers
- 22 hammer levers
- 23 hammer cushion
- 24 chime hammer lift wires
- 25 lower hammer assembly
- 26 hammers (4)

(20), with the chime warning pin (6) about a half turn away from the chime warning lever. This establishes the run of the pin at warning. Now finish putting the movement plates together. Leave the pillar screws finger tight, and set the remaining case brackets aside for later.

Install the locking plate (2) in a trial position, without pinning it on its post. Add the locking plate drive wheel (17) and fasten it with a taper pin. The movement may have index marks showing which of the two possible ways to put on the drive wheel, and where it is to mesh with the wheel under the locking plate. Naturally, you must still check the adjustment, to assure the correct chime note sequences.

Use the following method for establishing the note sequences. Turn the wheels manually, by touching the third wheel. The chime fly (7) should turn clockwise as you face the front of the movement. Watch the hammers rise and fall. Having determined beforehand which bell or rod sequence produces the four descending notes from the Westminster chime, look for that pattern in the hammer action. In some Sonora Chime clocks, the rods or bells are not physically located in descending note order within the clock case. After observing the hammer pattern you have been looking for, stop the gears at that point. Install the locking plate so the chime lock piece rests in the 1/4 hour locking plate slot. In Figure 43, the lock piece is in the 3/4 hour slot; the 1/4 hour slot is oriented at about 5:00 in the drawing. Visualize yourself pulling the locking plate out enough to disengage the locking plate drive wheel (17), then turning the locking plate until the 1/4 hour slot is under the lock piece. Push the locking plate back on to engage the wheels. Install the taper pin.

More adjustments can be made before you complete the assembly of the movement. Turn the wheels manually again and look for proper warning and locking action. The chime cam may be slightly out of position at the end of each chime sequence. If this is the case, the gear train will sometimes fail to

CHAPTER 5 - SETH THOMAS SONORA CHIME

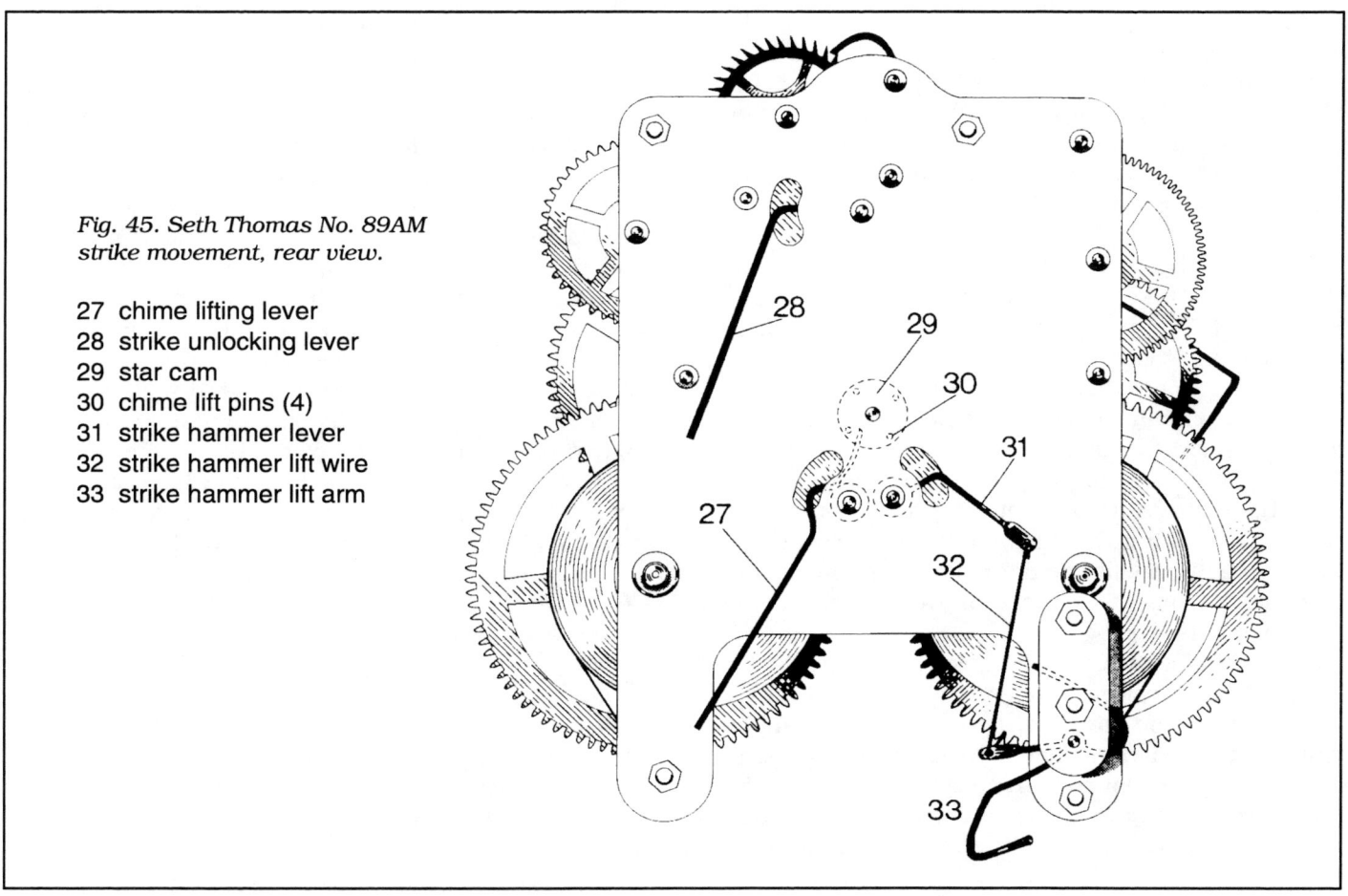

Fig. 45. Seth Thomas No. 89AM strike movement, rear view.

27 chime lifting lever
28 strike unlocking lever
29 star cam
30 chime lift pins (4)
31 strike hammer lever
32 strike hammer lift wire
33 strike hammer lift arm

lock. For locking to occur, the slot in the cam must be directly under the chime locking lever at the instant the chime lock piece falls into the slot in the locking plate. If an adjustment is necessary, separate the plates slightly to accomplish it. First, make sure there is no mainspring power on. Then separate the plates enough to ease out the front pivot of the locking wheel arbor, which carries the chime cam. Move the locking wheel and chime cam around to the locking position under the chime locking lever. Check for correct locking.

Observe the warning action. There should be about 1/2 revolution of the chime warning pin (6) before it hits the chime warning lever (8). Although you may have adjusted the warning as you put the plates together, it could easily have become upset again during the assembly and adjustment procedure. To correct the warning run, separate the plates and disengage the fly pinion, which carries the warning pin, from the locking wheel. Recheck after the adjustment.

You are now ready to finish assembling the rest of the chime movement. Install the click wheel (14) with the large washer and pin. Add the winding arbor (11), winding gear (12), and curved cover plate. Insert the pin to hold the chime lock piece, chime locking lever, and chime warning lever in place. Add the strike lifting lever (10). Install the chime silencer lever (19) on the rear of the movement. Add the upper case brackets, then tighten the pillar screws. Wind the mainspring and oil the movement.

No. 89 STRIKE MOVEMENT

The strike movement is as much a part of the Sonora Chime as the chime mechanism. It is therefore just as important to clean and repair the strike movement. Think of the No. 89 as the time and hour strike trains of a three-train chime clock. It just happens that these gear trains are in the form of a separate mechanism, instead of an integral part of a single chime movement.

Figure 45 shows the rear of the No. 89 strike movement used in the Sonora Chime clock with the No. 119 chime movement. The back plate is solid, with cutouts for the levers which connect it to the chime movement, and a cutout for the hammer lever. The pendulum suspension and regulator parts are omitted from the drawing. Figure 46 represents the front left side of the movement, with major strike parts.

It is a good idea to check for worn pivot holes before you take the movement apart. A brief inspection helps you uncover what needs to be done. The No. 89 is especially prone to excessive wear on the front pivot hole for the third strike arbor. This is the arbor which carries the cam (38). The escape arbor pivot holes are also likely to be worn. In fact, wear on the escape holes seems to be what stops many of the No. 89 movements after so many years of service.

To begin the repair, let the mainsprings down into clamps. Make sure the springs are completely let down. If the springs are sticky and dirty, they can fool you. Pressing on the clamp may separate some sticking coils, releasing more power. Let down this added tension before you go on. Next, look for the small spring studs on the movement plates. The brass wire springs for the strike levers are hooked onto these studs instead of wrapped around the pillars. Unhooking the springs now will lessen the chance that you will break or distort them as you separate the plates. Replace any damaged spring with a new one.

Separate the plates and remove all the parts. Do not forget to check the clicks. Many No. 89 movements have loose clicks, which you can repair by tightening the rivets. Check the click springs, and increase the tension if necessary. Remove the clamps from the mainsprings, using a spring winder.

Now clean and dry the movement. Install the required bushings, and polish any rough pivots. Reinstall the spring clamps. After a considerable amount of work, you will be ready to put the movement together again and adjust it.

Install all the arbors, including the main wheels and mainsprings, into the rear plate, then add the front plate. Carefully fit all the pivots into their holes. You may be able to mesh the strike train wheels correctly as you proceed, but it is difficult to do so as you juggle so many parts. You can adjust the parts after the plates are safely together.

Leave the mainsprings let down in the clamps. With finger pressure, turn the strike train to the locked position. With the count lever (37) in one of the deep slots in the count wheel (34), the drop lever (43) should be centered in one of the two slots in the cam (38). The strike lock pin (41) should be at the strike locking lever (42) as shown in Figure 46. Separate the plates and ease the arbor for the strike locking wheel (40) out of place to enable you to mesh the wheels correctly.

Once these adjustments are complete, the strike train should work. There may be several other adjustments required, however. Tighten the pillar nuts, then wind the mainsprings partially. You can wind the mainsprings until the clamps are loose, then

Fig. 46. No. 89 AM, strike train front view.

27 chime lifting lever
28 strike unlocking lever (outside rear plate)
29 star cam
34 count wheel
35 strike main wheel
36 count wheel drive pinion
37 count lever (cut to show drop lever)
38 cam
39 hammer lift pins (2)
40 strike locking wheel
41 strike lock pin
42 strike locking lever
43 drop lever
44 strike unlocking lever (between plates)
45 strike hammer tail
46 center arbor
47 strike fly
48 strike warning lever

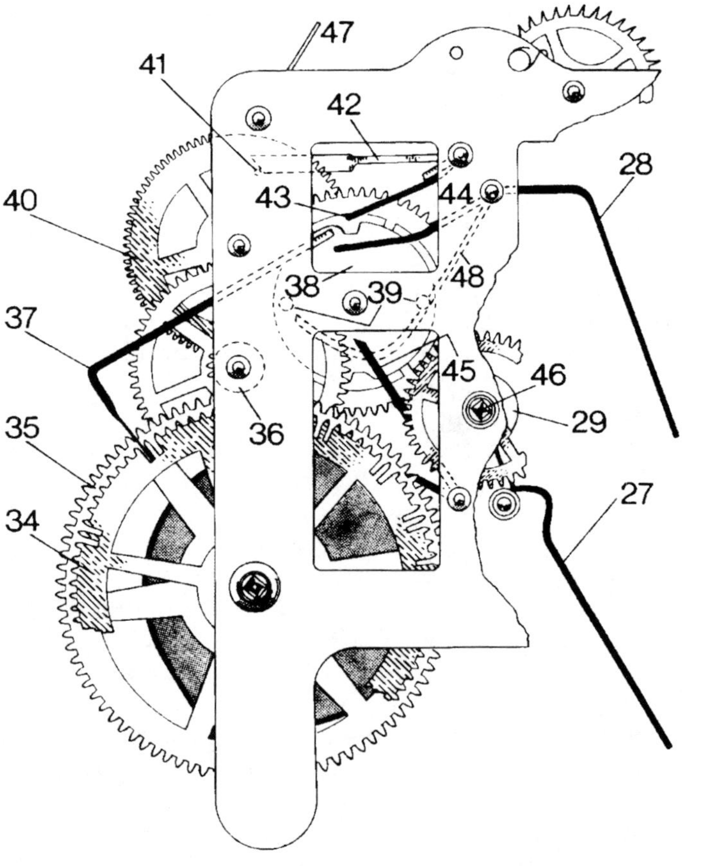

CHAPTER 5 - SETH THOMAS SONORA CHIME

remove the clamps. Push down on the strike unlocking lever (28) just enough to release the train. Now check the strike warning lever (48) to make sure that it catches the strike lock pin safely. If it does not, bend the warning lever upward to correct the problem.

Release the strike unlocking lever to permit striking to begin. Observe whether the count lever dips neatly into the center of each slot, especially the deep slot at the end of the strike sequence. Bend the end of the count lever to achieve the correct adjustment. The count lever should point at the center of the count wheel. That is the only way the lever can stay centered in the slots, which are radial. If the lever does not safely clear the tips of the count wheel during striking, bend the drop lever (43) downward to increase the height of the lever.

One of the most important adjustments is that of the strike locking lever (42). You may have to bend it up for shallower locking if the lock pin glances off the lever as the clock strikes. The opposite is required if the locking action is not deep enough. Bend the strike locking lever down for a deeper lock. If the strike train does not lock dependably at the end of each sequence, it will run on occasionally (or continuously) and the strike count will be wrong.

Another adjustment often overlooked is that of calibrating the minute hand so the No. 89 will release the separate chime movement as exactly as possible at each quarter hour. Some clocks have minute hands with movable bushings which permit adjustment of the hand. The Seth Thomas movement, like most American clocks, does not have this feature. The minute hand is one piece. Bending the hand will only distort or break it. Instead, bend the tail of the chime lifting lever (27), which operates with the pins on the star cam (29). There isn't much room for adjustment, but there probably won't be as much need for it as on other clocks with longer levers.

COORDINATING THE CHIME & STRIKE

With the two movements repaired to good working order, the task of installing them and getting them to work together remains. Installing them in the case may be a tedious process; consider some testing first. You could set up the No. 89 movement on a test stand or rack and run it for a few hours to check the time train. You could also spend a few minutes looking over the chime movement and watching the warning, locking, and hammer action. But there is no practical way to test the movements together outside the clock case. The case determines the relative positions of the movements, an all-important factor in how they will work together.

The No. 89 movement goes in the case first. Methods may vary according to the case, but one Sonora clock has the time and strike movement screwed to a steel plate. The plate is fastened to the case with wood screws.

As you install the chime movement behind the strike movement, look for the two levers coming from the back of the No. 89. Observe the upper lever, the strike unlocking lever (28). The strike lifting lever (10, Figure 43) on the chime movement must be placed to the left of it. Now check the lower lever, the chime lifting lever (27). The chime unlocking lever (9) is placed under and to the left of it. Actually, it's hard to get the levers wrong unless you force them. If you install the minute hand and turn it, you will soon see that the chime lifting lever must raise the chime unlocking lever if the chimes are to start. As the hour chime ends, the strike lifting lever must push down on the strike unlocking lever to release the strike train.

When you are certain the levers are in the correct positions, fasten the chime movement brackets to the clock case. It may take several attempts to get everything to fit into place. Wind the mainsprings enough so the gear trains will operate. Turn the minute hand to the next quarter hour. Check to see whether the chime movement unlocks. If it does not, bend the chime lifting lever (27) up so it will provide more lift. Look at the chime warning pin (6). You may need to bend it slightly to assure that it will hit the warning lever at warning, yet remain clear during chiming.

At the hour, watch the strike train. If it doesn't unlock, it is because the strike lifting lever (10) does not push the strike unlocking lever (28) far enough. Bend the strike unlocking lever to the left, closer to the strike lifting lever: this will produce higher lift.

Neither movement has a mechanism for correcting chime or strike automatically. To install the hands, put them on in accordance with the notes struck. Put on the minute hand, without pinning it. Turn through several quarter hours, listening to the chimes. After the hour chime and strike, simply install the hour hand pointing to the hour just struck. The minute hand goes on straight up. As you test the Sonora Chime clock, check for any discrepancy between the hand position and chimes or strike. It means there is a problem in one or both movements.

6
NEW HAVEN

New Haven is certainly one of the famous names in American clockmaking. The company began in the 1850's during the early years of the age of mass-produced clocks. At first, New Haven made thirty-hour brass works, but in time the firm expanded its line to include a variety of eight day lever and pendulum movements. In this chapter, we will look at the New Haven round chime movement. It is a common movement, but quite different from other chime mechanisms. The movement itself is not marked with a model number. However, the following list of patent dates is stamped on the rear plate of some examples:

> NEW HAVEN CLOCK C0.
> PATENTED
> JAN-1-1918
> FEB-19-1918
> NOV-28-1922
> AUG-14-1923
> DEC-25-1923

Throughout this chapter I will refer to the spring driven pendulum model New Haven. But I should mention that the company also offered an electric version marked "OBx" on the lower part of the rear plate. The electric motor runs the time train and also winds two small mainsprings to power the chime and strike trains. These can be wound up with an auxiliary key from the rear of the clock. Aside from the mode of power, it is the same as the pendulum movement for purposes of adjustment.

The New Haven pendulum movement is quite small at only 3-7/8 inches in diameter, or 4-5/8 inches including the mainspring barrels. The movement plays Westminster chimes on four rods. On the clocks with larger cases, the movement strikes the hour on a fifth rod. Smaller tambour models have room for only four rods altogether, so the movements are modified to strike on the first and third rods instead.

The time and chime mainsprings are the same size, 7/8 inch wide and .014 inch thick. A mainspring length formula was used to calculate the ideal length for these barrels as 64.6 inches. A 72 inch long mainspring is available from suppliers; it could be shortened to approximately 65 inches to assure the maximum number of turns will be produced from a full winding. The strike mainspring is 3/4 inch wide and .010 inch thick, and the length formula calls for a mainspring 81 inches long. Unfortunately, the closest available mainspring measures 3/4 inch wide, .009 inch thick, and just 72 inches long. This spring will produce at least one less barrel turn on a full winding, compared to an 81 inch long spring.

Figure 47 shows the New Haven chime movement. Because of its small size, the movement was installed in a variety of clocks. Figure 48 shows one of the better case styles, a nicely proportioned banjo chime clock. Mantel clocks with this movement were made in Gothic or cathedral designs. New Haven also used the movement in a number of large and small tambours. For example, the "Toledo" tambour is only 18 inches wide at the base and 5 inches deep, small indeed for a chime tambour.

CHIME TRAIN OPERATION

The chime parts on the front of the movement are drawn in Figure 49. The chime cycle begins with the warning at least five minutes before each quarter hour. One of the four chime lift pins (1) raises the chime warning lever (3) which in turn pushes up the chime locking lever (7). The chime lock pin (4) is moved out of the way of the chime locking arm and fly (5 & 6), releasing them to turn. After only 30° of rotation, the chime locking arm is stopped by the chime warning lever. This is the warning.

The rear movement view in Figure 50 shows the locking plate (24) with its four slots. During the warning, the chime lock piece (18) comes up out of

CHAPTER 6 - NEW HAVEN

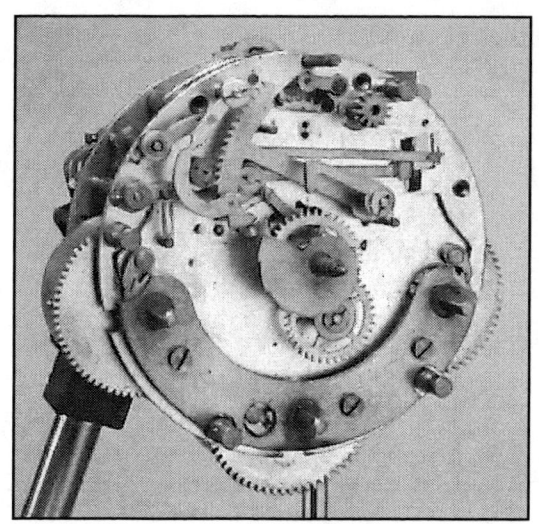

Fig. 47. The New Haven chime movement is unusual because of its round shape and small size. Also unusual is the placement of the barrels, with the time on the right and the chime in the center.

Fig. 48. The case is approximately 27" high including the eagle on top. There are few banjo clocks with Westminster chimes because of the space limitations of this style.

the slot. All the chime parts stay at their warning positions for several minutes until the quarter hour arrives.

Exactly at the quarter, the chime lift pin has moved just far enough around to release the chime warning lever. The lever drops down to its pre-warning position. Now the chime locking arm and fly are free to turn. The chime train starts running. The chime drive wheel (26), mounted on the locking plate, turns the pin barrel wheel (27). The pin barrel is made up of brass disks with tabs for lifting the hammers (29) in sequence to play the chime melody. Westminster chimes are counted as four notes at the first quarter, eight notes at the half hour, twelve notes at the third quarter, and sixteen notes on the hour.

Chime locking happens at the end of each chime note sequence. The chime lock piece drops down into the next slot in the locking plate. At the same time, the chime locking lever on the front of the movement drops, and the chime lock pin catches the spinning chime locking arm and fly. The gear train stops.

The New Haven movement automatically corrects the chimes if the sequence is upset. For example, someone may turn the minute hand either forward or backward so that the first quarter chime sounds at the half hour. Within one hour, the clock will correct any such chime error. Here's how it works. Look at the locking plate (24) in Figure 51 and you will see that its circumference is divided into four unequal arcs corresponding to the quarter hour chimes. The longest arc is for the 16-note hour chime. At the beginning of this arc there is a raised section. It takes a higher lift to raise the chime lock piece out of the slot and over the high spot to start the hour chime. Every quarter hour, one of the chime lift pins raises the chime linkage, but only one pin out of the four can lift high enough for the hour. This means the 16-note hour chime can sound only on the hour, and not at any other time.

STRIKE TRAIN OPERATION

The New Haven strike train is a rack and snail type. To find the start of the strike cycle, refer to Figures 50 and 51. There is a strike lift pin (25) driven into the underside of the locking plate. As the locking plate rotates through the hour chime, it brings the pin underneath the strike warning lever (22). The warning lever is pushed up, and it in turn raises the strike lifting lever (19). The strike locking arm (30) moves upward too, releasing the strike lock pin (28) which is on the locking wheel (17). The train now goes to warning as the strike warning pin (20), mounted on the warning wheel (21) is stopped by the strike warning lever.

Going back to the front of the movement and Figure 49, locate the strike unlocking lever (10). Dur-

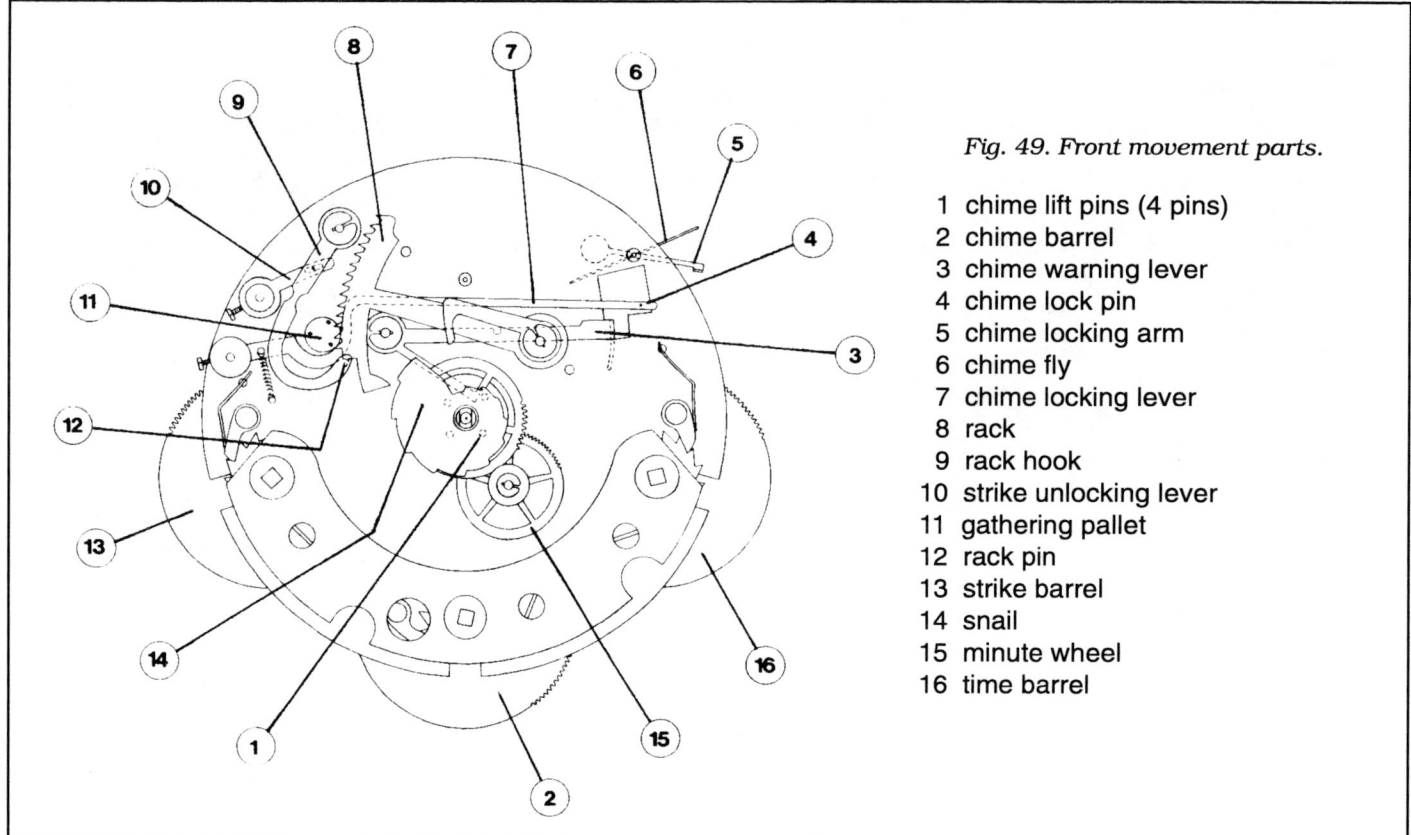

Fig. 49. Front movement parts.

1 chime lift pins (4 pins)
2 chime barrel
3 chime warning lever
4 chime lock pin
5 chime locking arm
6 chime fly
7 chime locking lever
8 rack
9 rack hook
10 strike unlocking lever
11 gathering pallet
12 rack pin
13 strike barrel
14 snail
15 minute wheel
16 time barrel

ing the warning, this lever lifts the rack hook (9). The rack (8) is released to drop onto the snail (14). Just as the hour chime ends, the strike lift pin has gone far enough to drop the strike warning lever. The strike warning pin now escapes from the warning lever and the train starts to run. The gathering pallet (11) counts off rack teeth to measure out the correct number of hammer blows for the hour. At the end of the count, the strike lock piece and the rack pin (12) reach the end of the rack. The strike locking arm (inside of movement) stops the strike lock pin. The strike cycle is finished.

REASSEMBLY AND ADJUSTMENT

Perhaps the most unusual feature of the New Haven movement is the overall layout. The company did not use a conventional design in which three gear trains are placed next to each other. Instead, a third movement plate was added between the other two. This allowed the three trains to be arranged in overlapping tiers within a compact space.

Figure 52 is a guide to the various train wheels and arbors. Use it to separate the parts into chime, strike, and time. There are three different lengths of arbors within the movement. Some run from the front to the rear plate, others from the front to the middle plate, and the rest from the middle to the rear plate. So you can use length to help you identify an arbor. Figure 53 is a further aid to the location of wheels. It shows the time train only, viewed from the rear, with the back plate off. The wheels drawn in dotted lines are under the center (third) movement plate (31), behind the front plate. Note that the time and chime barrels are reversed from their normal locations in most chime movements. This alone is the source of much confusion.

After assembling the movement, make sure the strike train locking and warning wheels are set with the pins arranged as in Figure 51. They should be assembled this way so the warning pin can run about a half revolution at the warning. If the locking wheel and the pinion on the warning arbor mesh incorrectly, the warning pin travel will be wrong each time. To adjust them, take off the chime lock piece (18) and the strike lifting lever (19). The mainspring barrels should be out of the movement to make the work easier, but they must in any case be fully let down. It is also a good idea to have the pendulum suspension unit out of the clock. Loosen the pillar screws and separate the plates enough to permit easing the strike train pivots out of their holes. Carefully mesh the strike locking wheel and the pinion on the warning arbor together the right way. The purpose of all this is to assure that you are going to have the correct strike warning. Clocks that have been improperly assembled with little or no run of the wheels at warning tend to stall when the strike should start up.

CHAPTER 6 - NEW HAVEN

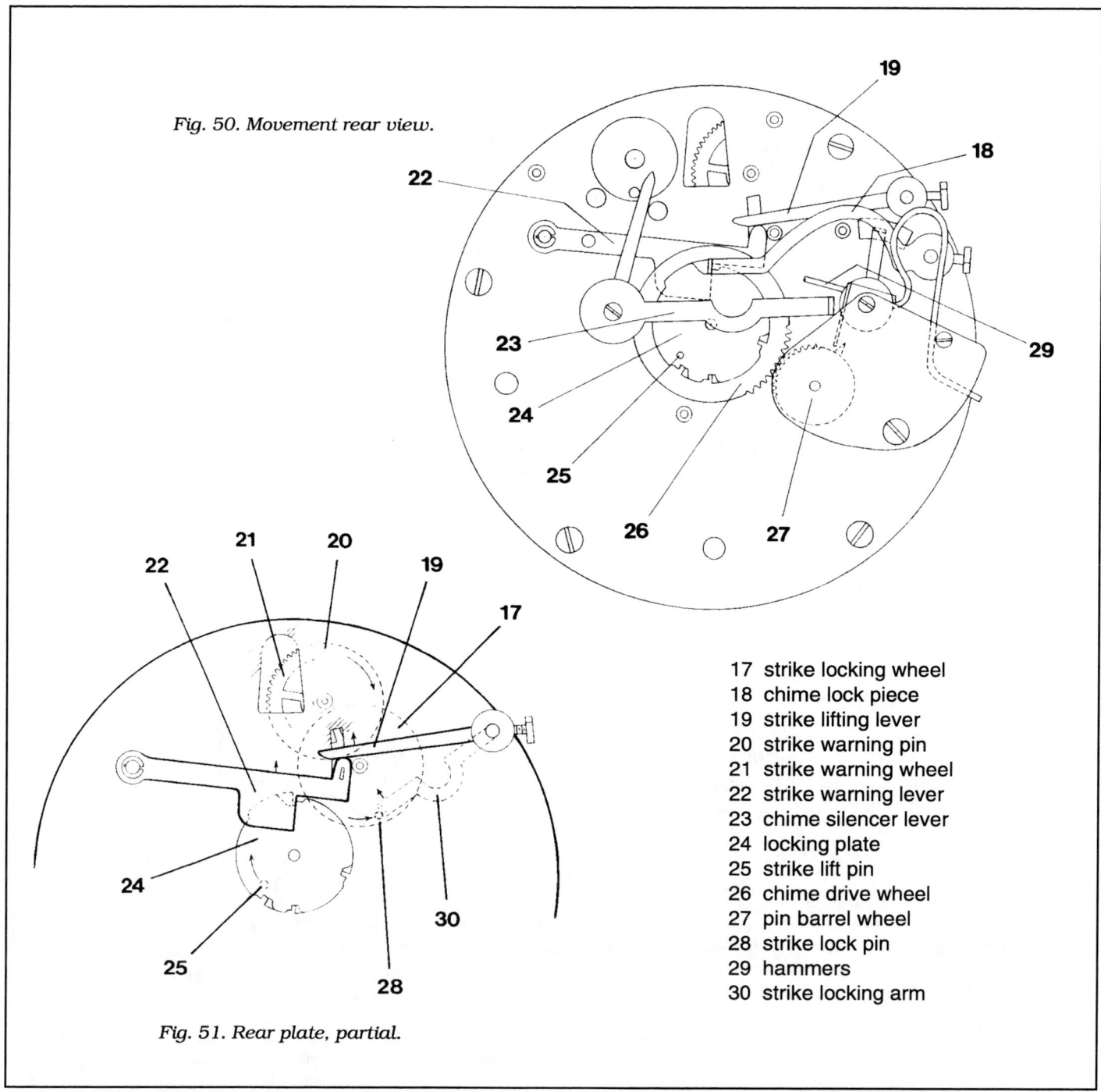

Fig. 50. Movement rear view.

Fig. 51. Rear plate, partial.

17 strike locking wheel
18 chime lock piece
19 strike lifting lever
20 strike warning pin
21 strike warning wheel
22 strike warning lever
23 chime silencer lever
24 locking plate
25 strike lift pin
26 chime drive wheel
27 pin barrel wheel
28 strike lock pin
29 hammers
30 strike locking arm

Finish assembling the movement and add the mainspring barrels. Wind partially, install the minute hand, and turn it through several quarter hours. Watch the wheels as you begin to make adjustments. Make sure that on the strike warning, the strike unlocking lever lifts the rack hook high enough to release the rack. If it does not, reset the strike lifting lever on the back of the movement so that it rests on the strike warning lever. This will increase the lift. The point of the adjustment is to assure release of the rack each hour.

Check the gathering pallet for correct setting. As the clock finishes striking the hour, the three pins on the gathering pallet must be clear of the rack as shown in Figure 49. If one of the pins remains engaged in the rack, the rack cannot be free to move down toward the snail at the next hour. To remedy the problem, carefully turn the gathering pallet to a new position on the elongated pivot which carries it. Don't bend or break the pivot; if the gathering pallet will not turn, it must be removed and then put back on at the correct place. When the gathering pallet pins are not touching a rack tooth at the end of the strike cycle, the adjustment is correct as

Fig. 52. Identifying the New Haven train wheels and arbors.

ARBOR	STRIKE	TIME	CHIME
Barrel	F.L. left side	F.L. right side	F.L. center
2nd	F.L.	F.L. pinion and wheel on same end of arbor	F.L. larger than strike second wheel
3rd	F.L. carries hammer lifting star	F. center arbor	R. carries long pivot for locking plate
4th	R. carries strike lock pin	F. pinion and wheel on opposite ends of arbor	R.
5th	R. carries strike warning pin	F. pinion and wheel on same end of arbor	R.
6th	R. fly	F. escape wheel	F.L. fly

KEY
F.L. Full length from front to rear plates, 35mm.
F. Arbor extending from front to middle plate, 11mm.
R. Arbor extending from middle plate to rear plate, 26mm.

Notes on identifying mainsprings and winding parts: The time barrel winds clockwise. The chime and strike wind counterclockwise. All three barrels have the cover at the rear, opposite the winding end. The strike barrel takes a narrower spring, and the design of the strike barrel arbor reflects this. To identify the time barrel arbor, note that it is made to permit the barrel to go furthest back of the three (due to time train second wheel). Look at the hooks on the three winding arbors, and you can see which way they wind by the orientation of the hook on each one. And remember, the time and chime barrels are reversed from the "normal" position they have in chime clocks.

far as the rack is concerned.

Look at the strike hammer to make sure it is not left hanging in the raised position at the end of the strike. If it is left "on the rise" you will have to go back and adjust the gathering pallet further. You will note from Figure 49 that there are three pins on the gathering pallet. Because of this, there are three ways to set the gathering pallet so the rack will not get caught on them when it is supposed to fall upon the snail at the hour. Try the other two positions, and one of them should result in the strike hammer finishing in the downward attitude.

Check the snail position at 12 and 1 o'clock to make sure the rack tall falls in the correct positions at the hour. If the snail must be moved, first remove the large washer over the minute wheel (15). Then pull the hour wheel and snail away from the pinion. Reengage at the correct point.

CHIME ADJUSTMENTS

In this movement, the chime adjustments are critical and must be done carefully. At the upper right front area of the movement, at about a 2 o'clock orientation, is the area we will study closely for chime adjustments. The chime warning and locking levers enter the movement at this point, to act upon the chime locking arm. Clearances are small.

CHAPTER 6 - NEW HAVEN

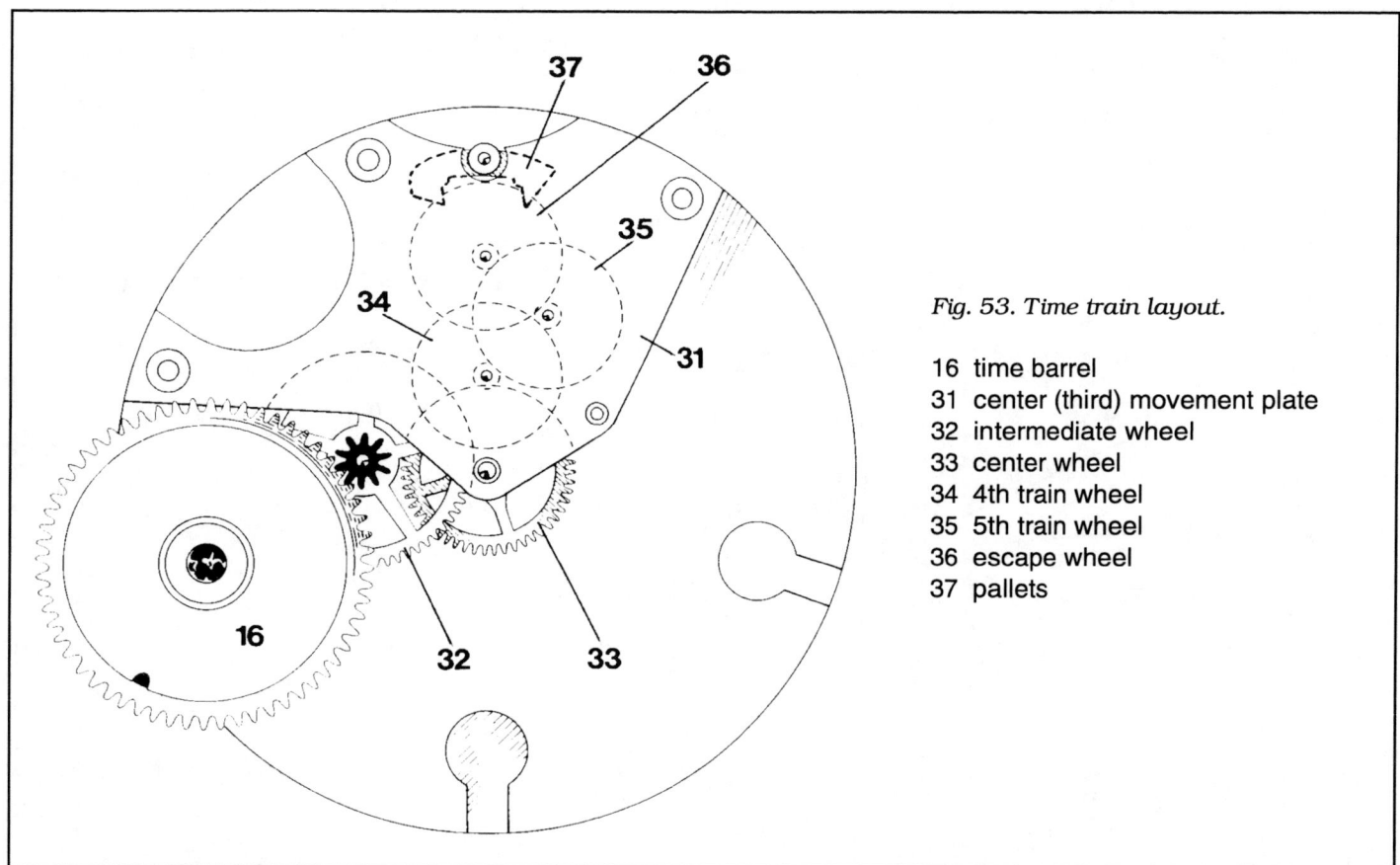

Fig. 53. Time train layout.

16 time barrel
31 center (third) movement plate
32 intermediate wheel
33 center wheel
34 4th train wheel
35 5th train wheel
36 escape wheel
37 pallets

Unfortunately, the means for making the necessary adjustments are not so fine: you have to work with crude square-headed set screws.

The chime locking lever (7) reaches across the front of the movement. It is fastened to its arbor by a set screw. Often, the screw is turned down very tightly, and the screw digs into the arbor. Fine adjustments are difficult because the lever may move as you tighten the screw. Turn the screw firmly, however, or the setting will wander off, causing a chime problem. The chime locking lever is a weak point in the clock. Often a problem encountered during testing will be found to be related to it.

Before attempting adjustments, pick up the movement and review how the chime train works. Find the locking plate on the back of the movement. The chime lock piece rides on it, to determine the length of each chime sequence. The chime lock piece is fastened to the same arbor which carries the chime locking lever on the front of the movement. It is also held in place by a set screw. Thus the two levers, one on the front of the movement and the other on the back, form a complete unit. Their precise adjustment is necessary for the chime train to function. There is no margin for error.

Figures 54, 55, and 56 show the area on the front of the movement where chime warning and locking occur. We will look at the three positions of the levers, for lock, warning, and run. These are shown in the three drawings. If the adjustment is correct for all three positions, the movement will chime. To observe the chime action, hold the movement in your hand and turn the minute hand slowly. Look for each position. You will need to watch the front movement area shown in Figures 54 through 56 and the back of the movement as well.

Lock position: the chime locking arm (see detail in Figure 57) is carried on the fly arbor. Although the arm is driven tightly on the arbor, the fly is not. As a result, the relative positions of the fly and the arm do not remain constant. Figure 54 shows the lock position, with the chime locking arm stopped against the chime lock pin (4). At the same time, the chime lock piece on the back of the movement will rest in one of the four slots in the locking plate.

Warning position: before each quarter hour, the chime warning lever (3) raises the chime locking lever (7). Figure 55 shows that the chime lock pin has moved up, permitting the chime locking arm to proceed in a clockwise direction. It moves only the short distance to the warning lever. This warning action is fixed, and cannot be adjusted. The chime locking arm runs from an established point, the chime lock pin. It does not have to be set.

Run position: exactly at each quarter, the chime warning lever drops to release the chime locking

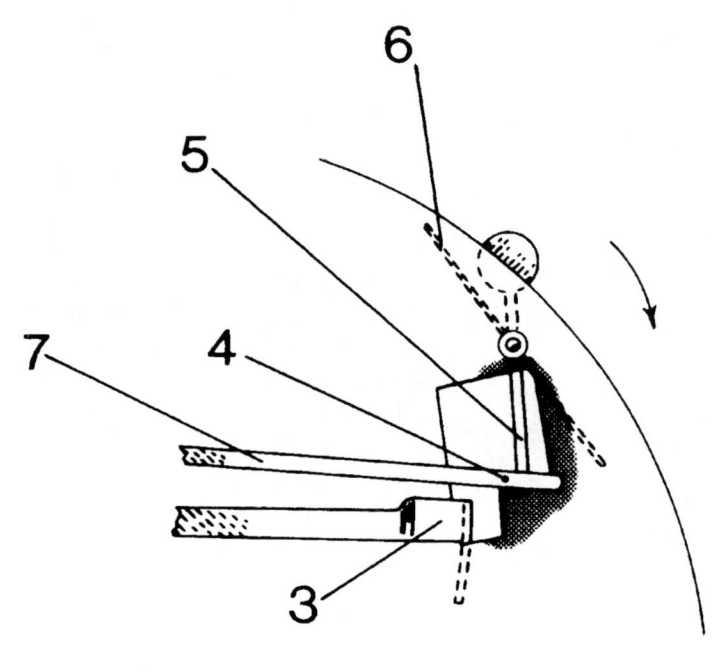

Fig. 54. Front detail showing chime locking.

3 chime warning lever
4 chime lock pin
5 chime locking arm
6 chime fly
7 chime locking lever
38 center arbor

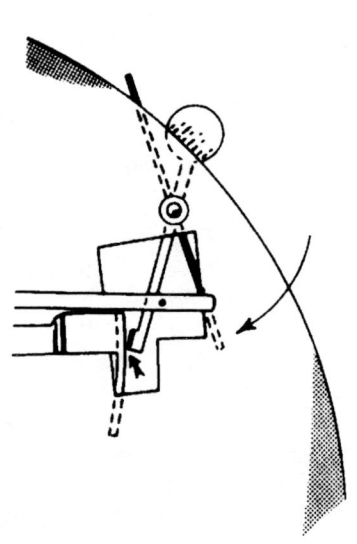

Fig. 55. Chime warning position.

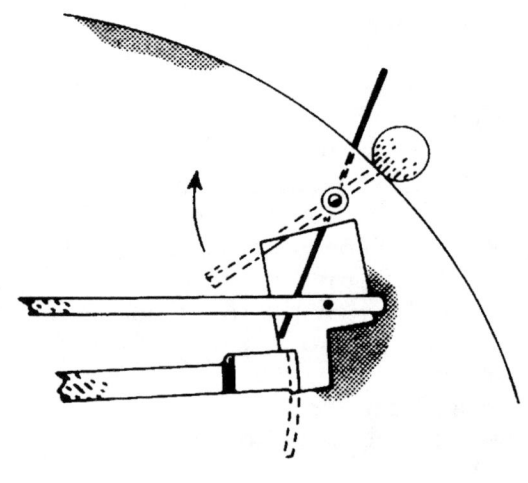

Fig. 56. Chime running position.

CHAPTER 6 - NEW HAVEN

arm. Figure 56 shows that the tip of the chime locking arm runs below the chime lock pin and above the chime warning lever. The chime train runs until the chime lock piece reaches the next slot in the locking plate. At this time the chime locking lever and pin move down. When the pin blocks the chime locking arm, the gear train stops.

Assuming the chime train will run, the critical time to watch is the moment before chiming ends. A small step is cut in the locking plate just before each slot. As the chime lock piece enters a step, it moves part of the way down toward the lock position. The "run" clearance between the chime lock pin (4) and the chime locking arm (5) is reduced to a minimum. The pin is poised to stop the arm at the instant it drops.

Often, the chime lock pin hits the arm too early, when the chime lock piece is on the "step". This happens because the chime locking lever is too low. To correct the problem, set the chime lock piece on the back of the movement for a higher lift. Push the lever down to accomplish the change. Loosen the set screw just enough to allow movement, then tighten again.

The opposite problem occurs if you have overdone the adjustment by pushing the chime lock piece down to provide more lift. After this adjustment, it also requires more lift to bring the chime lock piece up out of the slot in the locking plate. The net result may be that nothing happens. The warning lever does not push the chime locking lever high enough to raise the chime lock piece. Remember, the chime locking lever on the front of the movement and the chime lock piece on the back are fastened to the same arbor. Whatever you do to change one lever affects the other.

The adjustment procedure may end up as a juggling act between these two extremes. Either the chime train doesn't unlock at all, or it locks too soon. Only careful adjustments will assure good results. Just tightening the set screws seems to change the adjustment. It is probably better to leave the set screw on the front of the movement alone, concentrating instead on the set screw holding the chime lock piece on the back of the movement. When you feel the adjustment is correct, turn the minute hand to make the movement go to the warning position, then the run position. Stop the fly at critical points so you can observe the clearance between the chime lock pin and locking arm. If you are satisfied, double check the tightness of the set screws.

If all attempts at chime adjustment still result in

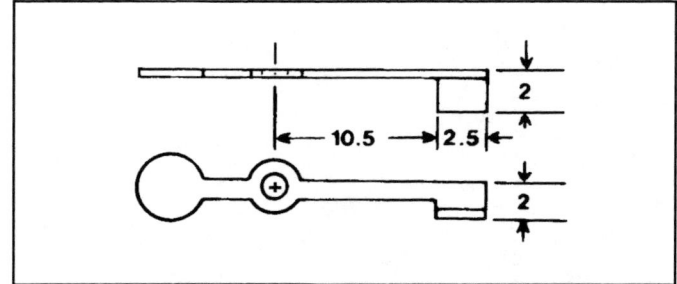

Fig. 57. Detail of chime locking arm, part #5. This stamped brass part is carried on the fly arbor. Dimensions are in millimeters. The chime locking arm is also shown in Figures 49 and 54-56.

either failure to release or failure to lock, check the chime warning lever (3). There should be a notch halfway along its length (see Figure 49). This notch allows the lever to be bent. If the right-hand side of the lever is bent upward slightly, the lifting action is increased. A preexisting downward bend might be the reason the chime train could not be adjusted.

Another common chime problem involves the spring-loaded disk behind the locking plate. In Figure 51 you can see a small portion of this disk showing in each of the four slots in the locking plate. The disk is supposed to slide underneath the chime lock piece as it comes up for the warning. If the disk does not move underneath the lock piece to support it, the piece just falls back into the same slot when the chimes are to begin. The clock does not chime. To correct the problem, all you usually need to do is to clean the locking plate and disk assembly to free up the spring action. If this is not the answer, look for an unhooked or damaged spring.

When you have the chime train working properly, check the hammer sequence. The easiest way with any Westminster chime clock is to listen for four descending notes at the first quarter. If a movement is out of its case, you can watch the hammers instead. First quarter chimes on the New Haven movement will be the four hammers rising and falling in order from front to rear. If the first quarter chime plays the right notes, the other quarters will also be correct. If the notes are wrong, or if you have a hammer left hanging in the raised position at the end of each sequence, adjustment is simple. Loosen the bracket plate which holds the pin barrel and hammers in place. (Refer to Figure 50.) Move the pin barrel wheel (27) away from the chime drive wheel. Using a trial and error method, reengage the wheels at the point where the chime notes are correct.

7
SESSIONS TWO-TRAIN CHIME

The Sessions chime movement we'll work on in this chapter is certainly not rare. It shows up in tambour cases and other mantel clock styles. I have included the Sessions movement in this book because it is rather different from other chime movements. This makes it more difficult to "learn" the Sessions chime movement. After you have restored one, you may forget many of the details of this unique movement before the next one comes along. It is essential to understand how the movement works before you start to change things. Accordingly, we will begin with an explanation of how the parts operate. Then we will cover a procedure for repairing and adjusting the movement.

The original patent for the Sessions chime-strike mechanism was dated December 22, 1931, as No. 1,837,462. The following year the inventor, Samuel Mazur of Bristol, Connecticut, filed another patent to incorporate certain improvements into the design. The second patent was No. 1,883,387, dated October 18, 1932. Figures 58 and 59 show front and back views from the patent drawings, including the "player discs" and dual rack system. The second patent states that its purpose was to "improve the counting mechanism" and to "improve the means for supporting part of the train of gearing which operates the chiming and striking mechanism so as to facilitate assembling of the same and also prevent distortion and binding .." A number of other changes were listed.

The whole idea of the combined chime-strike train was to produce a totally self-correcting mechanism in a two-train movement. In this sense the Sessions movement was a complete success. The pin barrel can be installed without regard for chime note sequences; it regulates itself. But aside from this, the Sessions can be a nightmare for the repairer.

The time mainspring is a loop-end type which is 3/4 inch wide, .018 inch thick, and 96 inches long. Many repairers use slightly thinner mainsprings in the .016-.017 inch range in place of the original "standard" springs. This reduces wear and deformation of the main wheel teeth. The chime-strike mainspring is 3/4 inch wide, .018 inch thick, and longer than the standard spring at 108 inches.

The movement has only two mainsprings and gear trains instead of the usual three to handle the chime, time, and strike. The Sessions is an "add-on" design, in which an American striking movement has been adapted to play Westminster chimes. Viewed from the front as in Figure 58, the time train is on the right side, and the chime-strike train is on the left. Each is powered by a loop-end mainspring. The movement plates are elongated to accommodate the large chime and strike wheels. The removable brackets shown in Figure 58 carry the front pivot holes for the winding arbors. If the mainsprings are first let down into wire loops, the brackets can be taken off. The main wheels and mainsprings can then be taken out without disassembly of the entire movement. The clock plays Westminster chimes on four rods, and strikes the hours on a fifth rod. A pin barrel made up of five flat disks, shown in Figure 59, lifts the hammers. A dual rack and snail arrangement controls the counting of the notes.

Figure 60 shows the major chime-strike parts except for the pin barrel. When it is not operating the hammers, the gear train is held in the locked position by the drop lever (10), which rests in one of the two slots in the locking cam (1). The star cam (6) is located on the center arbor (5) between the movement plates.

Before each quarter hour, the star cam raises the lifting lever (7). The lower unlocking lever (14) in turn raises the rack hook (13). The rack hook pushes up the upper unlocking lever (12), which is attached to the same arbor as the drop lever. As the drop lever comes out of the slot in the locking cam, the gear train is released for the warning. The warning pin and wheel move around until the warning lever (8) stops the pin. Observe that the warning lever is on the same arbor as the lifting lever. Ex-

CHAPTER 7 - SESSIONS TWO-TRAIN CHIME

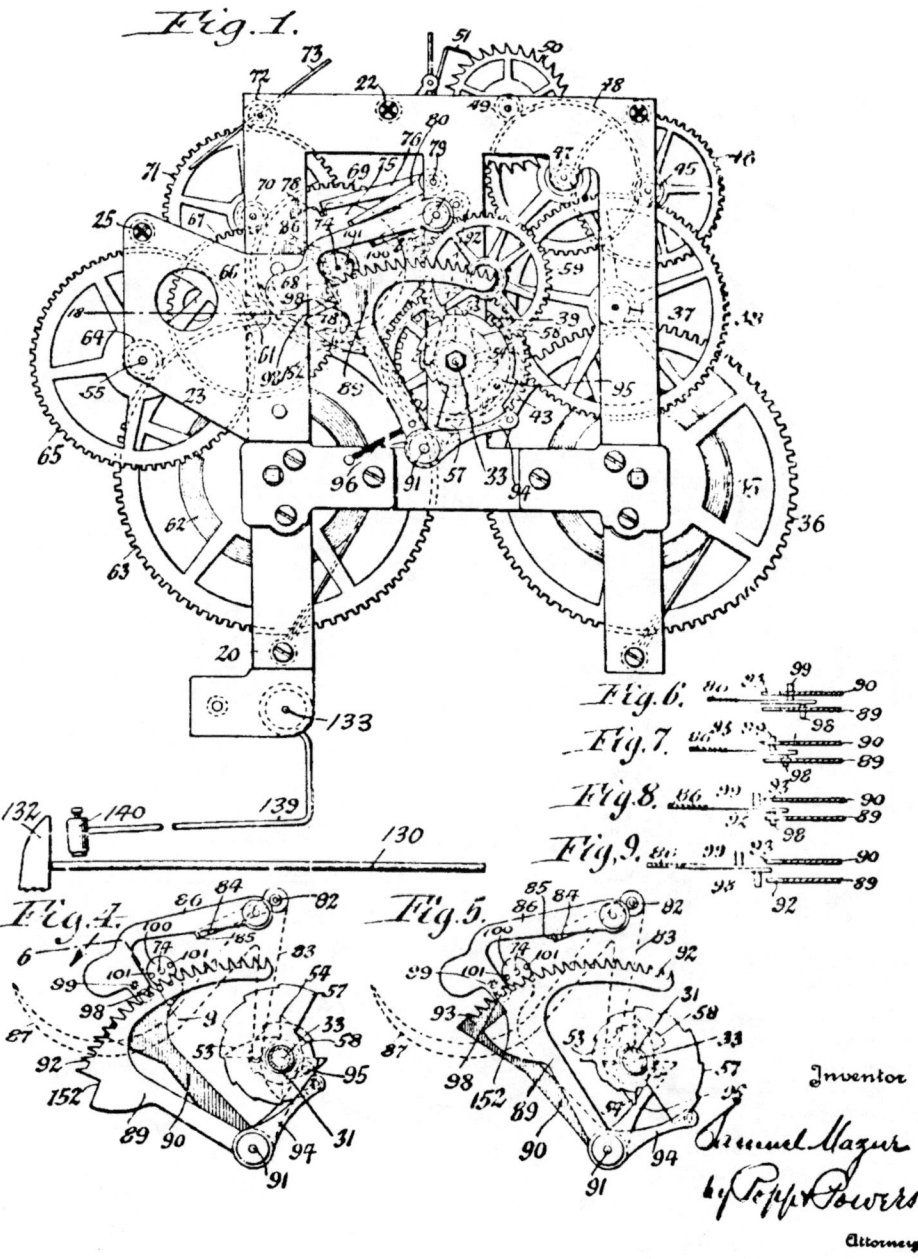

Fig. 58. Patent drawing showing front view of Sessions two-train chime movement and details of dual racks.

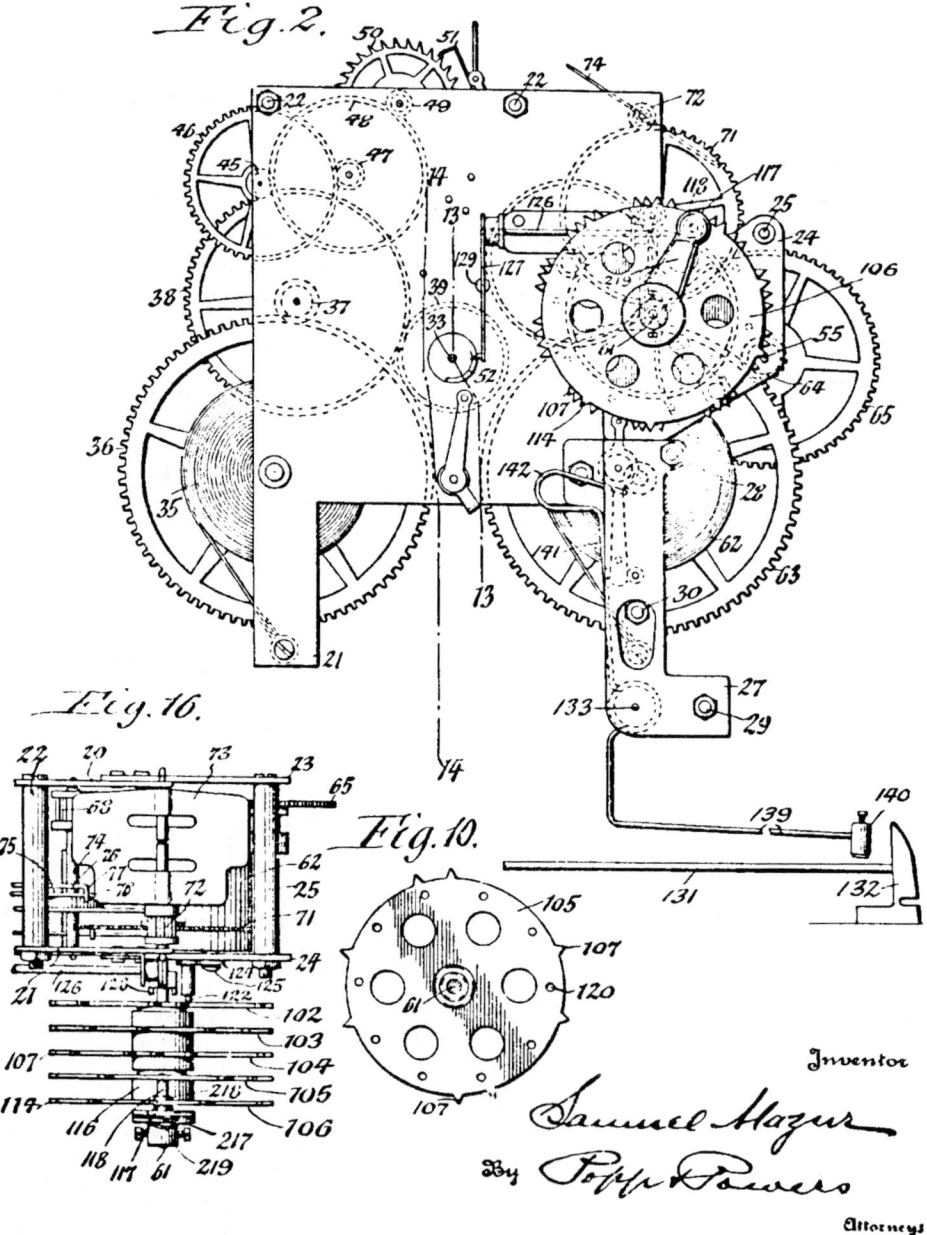

Fig. 59. Patent drawing of the rear of the movement shows the location of the pin barrel made up of "player discs".

CHAPTER 7 - SESSIONS TWO-TRAIN CHIME

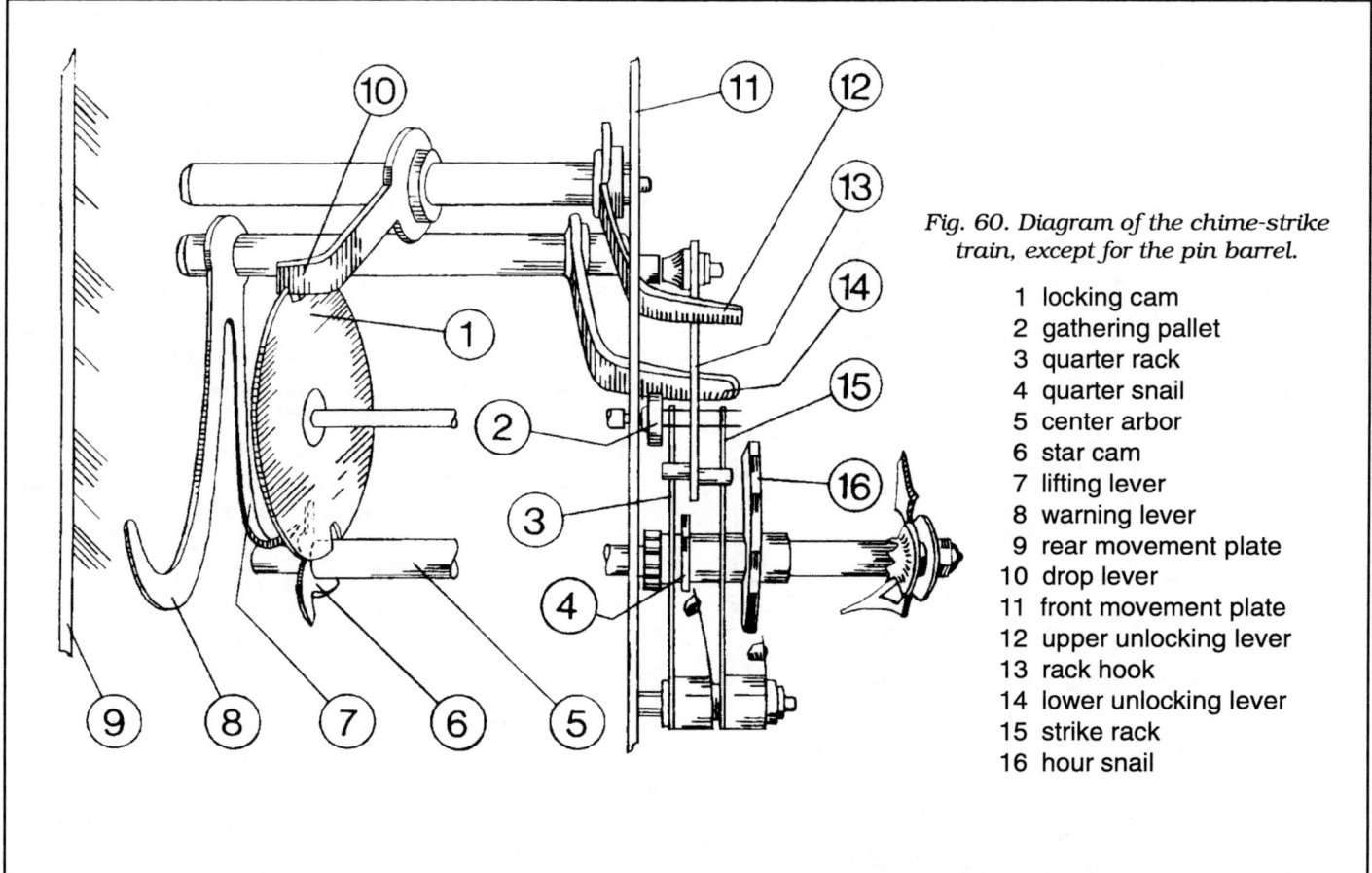

Fig. 60. Diagram of the chime-strike train, except for the pin barrel.

1 locking cam
2 gathering pallet
3 quarter rack
4 quarter snail
5 center arbor
6 star cam
7 lifting lever
8 warning lever
9 rear movement plate
10 drop lever
11 front movement plate
12 upper unlocking lever
13 rack hook
14 lower unlocking lever
15 strike rack
16 hour snail

actly at the quarter hour, the lifting lever is released by the star cam, allowing the warning lever to let go of the warning pin. The gear train runs as the two racks (3 and 15) count off the note sequences. At the end of the chiming and striking, the drop lever goes back into one of the slots in the locking cam. This occurs as the gathering pallet (2) counts off the last tooth of the quarter rack, which permits the rack hook to descend.

TWO RACKS

The quarter rack (3) is placed behind the strike rack (15), with the rack hook in between them. The quarter rack acts upon the quarter snail (4), which has one position for each of the quarter hours. For the first quarter, the rack tail falls to the first position on the snail. For the second quarter it falls further to the next position, and so on. The longest fall is for the hour chime, the longest note sequence. Figure 61 shows the quarter rack and snail in the hour position.

When you consider that one rack is for the strike and the other for the quarters, the idea seems simple enough. But it is easy to become confused if you watch the movement as it works. At the hour, chime note counting always begins on the strike rack. Then, up to four strike notes are counted on the quarter rack. Although this may seem backwards, it works. What is important is that the two racks count off the correct total number of notes. The pin barrel takes care of switching from chime to strike automatically. Examples will help explain.

9 O'clock Example

At 9 o'clock, the quarter rack goes to its "4" position, which it does each hour to allow for four rack teeth times four notes per tooth, to equal 16 hour chime notes. The strike rack goes to the "9" position for 9 o'clock. As the movement begins chiming, the gathering pallet counts on the strike rack first. The reason for this is seen in Figure 60. Note that, of the two rack pins which come out of the rack hook (13), one is for the quarter rack and the other is for the strike rack. The strike rack pin is set lower, so it engages the rack teeth. The quarter rack pin is higher. It does not rest upon the quarter rack until the strike rack is counted out and the rack pin drops off the end of the rack. Then the quarter rack pin can rest on the quarter rack and hold each tooth as the next one is gathered.

Continuing with our 9 o'clock example, counting for the hour chime begins on the strike rack as I have explained. Four strike rack teeth are gathered to count the hour chime. Then the clock auto-

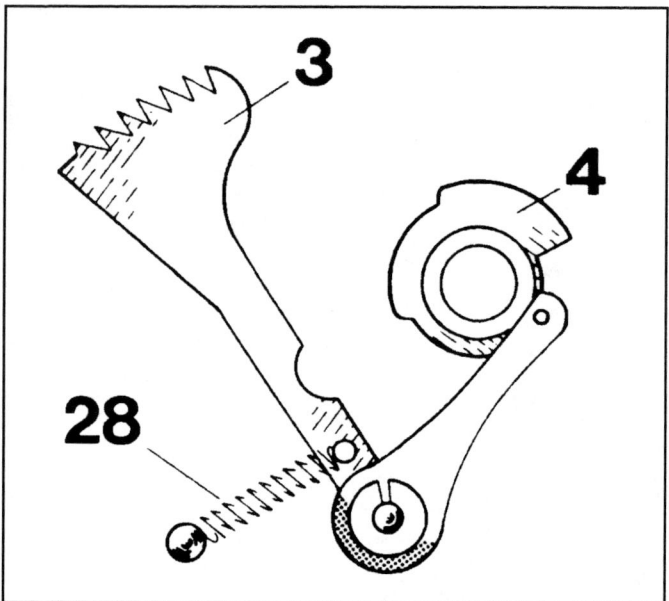

*Fig. 61. Quarter rack and snail at the hour position: **3** quarter rack; **4** quarter snail; **28** rack spring.*

matically begins counting the nine hour strike notes, while still on the strike rack. There are only five teeth left, however, because the hour chiming has used up the first four teeth. This means the first four notes must be counted from the quarter rack. A total of 13 teeth are gathered: first, nine teeth from the strike rack, and then four from the quarter rack.

1 O'clock Example
Another useful example is 1 o'clock. The quarter rack is at the "4" position as it is every hour. The strike rack fails to the "1" position for 1 o'clock. Hour chiming requires four teeth for 16 notes. For this it takes the single strike rack tooth, followed by three of the four quarter rack teeth. As the gear train continues to run, the pin barrel shifts automatically from chime to strike. The remaining quarter rack tooth is for the single hour strike note for 1 o'clock.

PIN BARREL
Now that we have covered the dual rack system, we know how the movement counts the correct number of chime and strike notes using only one gear train. The other major area we need to investigate is the pin barrel. It is made up of five disks which lift the chime and strike hammers. Figure 62 shows the arrangement of the disks. The strike disk is at the left, and the four chime disks are to the right in each view. The overall rear movement drawing in Figure 59 shows the orientation of the pin barrel on the rear of the movement, plus a view of a single disk. The disks are carried on the elongated third arbor of the chime-strike train.

The four chime disks are fastened as one unit, allowing the strike disk to operate independently of the others. The pin barrel unit is not set screwed onto its arbor. Instead, it is moved by the drive arm and pin. The arm is shown in Figures 59 and 62 on the rear of the strike disk, and it is represented as part (21) in Figure 62. The pin barrel switches automatically from chime to strike by sliding forward and back again at the right moments. The strike disk always rotates when the gear train is running for chime or strike, and it lines up to operate its hammer during the hour strike. The four-disk pin barrel unit rotates only during chiming.

Figure 62 is a side view of the pin barrel from the chime side of the movement, showing its three modes of operation. In these side views, "left" means to the rear of the clock and "right" means forward. The three positions are: "a" chiming second, third, and fourth quarters; "b" hour striking; and "c" chiming the first quarter.

Chiming 2nd, 3rd, 4th Quarters
In view a), the locking stud 123) holds the entire pin barrel unit to the left under spring pressure. Although the strike disk rotates, it is not lined up with the strike hammer (the hammer is not shown here), and striking does not occur. The drive pin (22) moves the whole unit because it extends through the strike disk (17) and also the rear chime disk (18). The four chime disks are fastened together, so they all move as a unit.

Hour Strike
At the hour, the pin barrel operates as described above until the last of the 16 chime notes is played. At that moment, the single hole in the chime disk comes into position over the locking stud. The pin barrel unit moves forward (to the right in the drawing) as it locks over the stud. The pin barrel now appears as in view "b". Two changes have occurred. First, the drive pin is no longer in contact with the chime disks, which are locked on the stud. Second, the strike disk has moved forward, in line with its hammer. The drive pin still pushes through the hole in the strike disk, but not through any of the chime disks. In "b" the hour is struck with only the strike disk rotating. Just after each hammer blow, the drive pin passes over one of the ten concentric holes in the rear chime disk. The strike sequence always ends with the pin positioned over (not through) one of these holes.

The chime correction feature built into this movement is quite simple. The two racks, as I have explained, count out the correct total number of chime

CHAPTER 7 - SESSIONS TWO-TRAIN CHIME

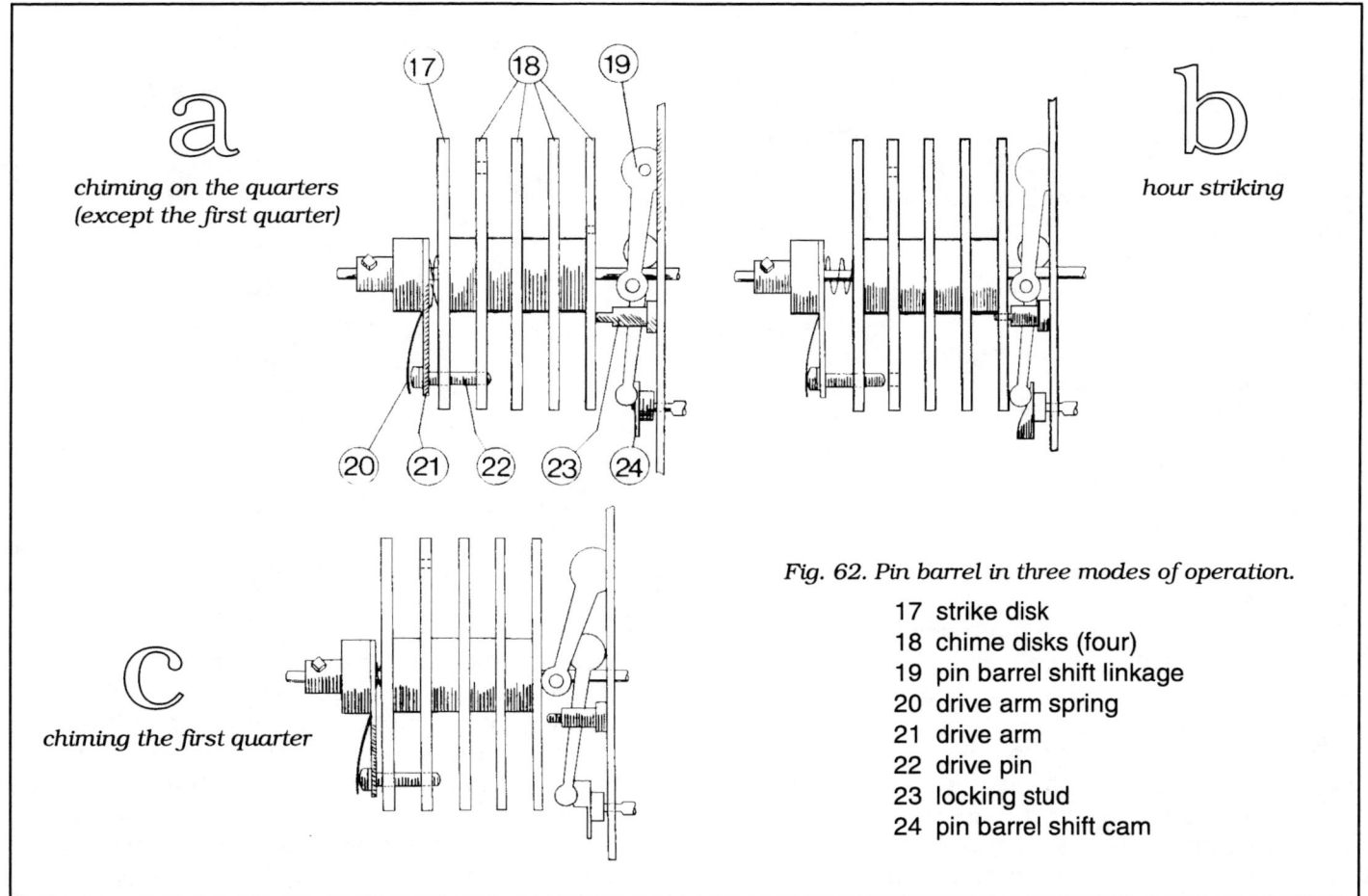

Fig. 62. Pin barrel in three modes of operation.

17 strike disk
18 chime disks (four)
19 pin barrel shift linkage
20 drive arm spring
21 drive arm
22 drive pin
23 locking stud
24 pin barrel shift cam

and strike notes. The pin barrel assembly automatically moves from chime to strike each hour. If the chimes get out of sequence for any reason, the correction is accomplished at the hour. Once the clock has started to chime the hour notes, it just continues to chime until the hole in the first disk comes around to the locking stud again. There is only one hole in this disk, and it marks the end of the 16-note hour tune, no matter how many Westminster chime notes have actually been played during the correction phase. Then the movement switches to strike and runs until both racks are counted out. At the first quarter, the clock will then play the correct four notes.

First Quarter

In view "c" from Figure 62 the pin barrel shift cam (24) and shift linkage (19) come into play. Gradually during the first 15 minutes of the hour, the cam (on the end of the center arbor) pushes on the linkage. Figure 63 shows this in progress. By the first quarter hour chime point, the cam and linkage have moved the pin barrel all the way to the rear (left in Figure 62, view "c"). This brings us into chiming position again. The pin barrel unit is now off the locking stud, the drive pin is sticking through the last chime disk, and the strike disk is no longer in line with its hammer. Chiming occurs. Then, at about 20 minutes past the hour, the cam drops the linkage, allowing spring pressure to push the pin barrel unit slightly forward again, to the "a" position. The locking stud holds the pin barrel in the chiming mode until the end of the next hour chime. Then we move to "b" in the diagram, and so on.

REPAIR AND ADJUSTMENT

As with any loop-end mainsprings, you must place a clamp or retainer on each one before using a let-down key to release the tension. On the chime-strike side, there may not be enough clearance to allow installation of the usual wire loops. The hammer assembly is in the way. Just remove the hammer assembly first, so you can gain access to the spring. Don't worry about mixing up the chime by taking off the hammer assembly. The mechanism automatically synchronizes itself as the movement chimes.

One particularly rough task is the removal of a broken chime-strike mainspring. I've encountered a few of these where the mainspring was shattered in many places. It wasn't possible to pull out the broken coils because of inadequate clearance. Attempts to let the spring down were not successful because of interference from the movement pillar and other arbors. The best procedure was to carefully apply heat to the spring with a torch, then cut through the softened coils with a hacksaw. This worked just fine; I was able to pull out the pieces. Always use heavy gloves when handling mainsprings which may have sharp edges.

With the mainsprings safely let down, remove the main wheels and springs before separating the plates. The brackets supporting the winding arbors are screwed in place, permitting easy removal. The lower movement pillars, held at each end by screws, can also come out, allowing you to pull out the main wheels. Double check to see that you have unhooked the two rack springs before taking off the bracket on the left.

There is one more obstacle to overcome before you can separate the plates. The silencer arbor is located at the bottom center of the movement, running between the plates. Pull off the front collar and tension washer. Now you can separate the plates and remove the wheels.

You may want to leave the pin barrel shift cam (24) on the back end of the center arbor. Figure 63 shows what the cam looks like. The center arbor will remain with the back plate if you leave the cam in place. If you do remove the cam, it is only necessary to put it back in the right orientation. We'll cover that later in the chapter. To remove the pin barrel and strike disk, loosen the set screws on the collar at the end of the arbor and take off the unit.

After cleaning the movement, you will have to take care of some bushing and pivot polishing work. The pivots seem to be very rough in these movements. The chime-strike train turns rather slowly as it is, so you should do your best with the repair. This does not mean you have to put a bushing in every hole, but you do have to correct any trouble spots. The reassembly job is not as hard as it looks. Just go one step at a time. Begin by installing the wheels and getting the plates together. The main wheels and mainsprings can be left out until later.

The first adjustment is represented in Figure 64. The locking cam (1) and drop lever (10) hold the chime-strike levers in the locked position. There is no separate locking pin as in many strike movements. Turn the wheels until the drop lever falls into one of the two slots in the locking cam.

Now observe the warning pin (26). It should be at about a 12 o'clock orientation. This allows for almost a half revolution of the warning wheel (25). The warning lever is not shown in the figure, but is located behind the wheels. If necessary, separate the plates just enough to permit adjustment of the warning wheel.

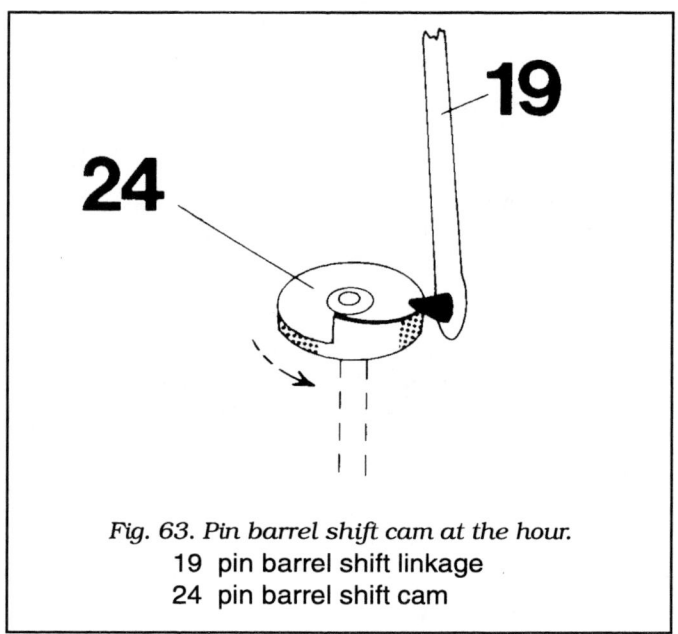

Fig. 63. Pin barrel shift cam at the hour.
19 pin barrel shift linkage
24 pin barrel shift cam

Next, install the minute hand and turn clockwise. The star cam, located on the center arbor between the plates, will lift the chime-strike levers at each 15-minute mark. Look for the higher lift, which is the hour. Now install the minute hand in the 12 o'clock position.

Figure 61 shows the quarter rack (3) and quarter snail (4). These are part of the dual rack and snail system. They are located behind the strike rack and hour snail in a completed movement. As you have just established the hour point with the minute hand, you need to install the quarter snail at the hour location as shown. There are four segments on the quarter snail, corresponding to the quarters. Install the snail to allow the rack tail to fall to the lowest segment of the snail. This is the hour chime setting. You can now add the hour snail to the hour pipe, and fasten it with the set screw. Before tightening, adjust the hour snail so that the strike rack tail falls directly to the 12 o'clock position. Check this carefully.

Turn to the back of the movement and Figure 63. If you left the pin barrel shift cam (24) on during repair, no adjustment is needed. If you have to put it back on the end of the center arbor, you must do it correctly now. The figure shows the cam as it is set following the end of the hour chime and strike. This can be confusing. The cam does its work during the interval after the hour, and before the first quarter. Its function is to move the pin barrel off the locking stud, to change the musical "mode" from strike back to chime. To anyone trying to understand this movement for the first time, it would seem logical that the cam has something to do with setting up the mechanism to chime the hour. That is

CHAPTER 7 - SESSIONS TWO-TRAIN CHIME 51

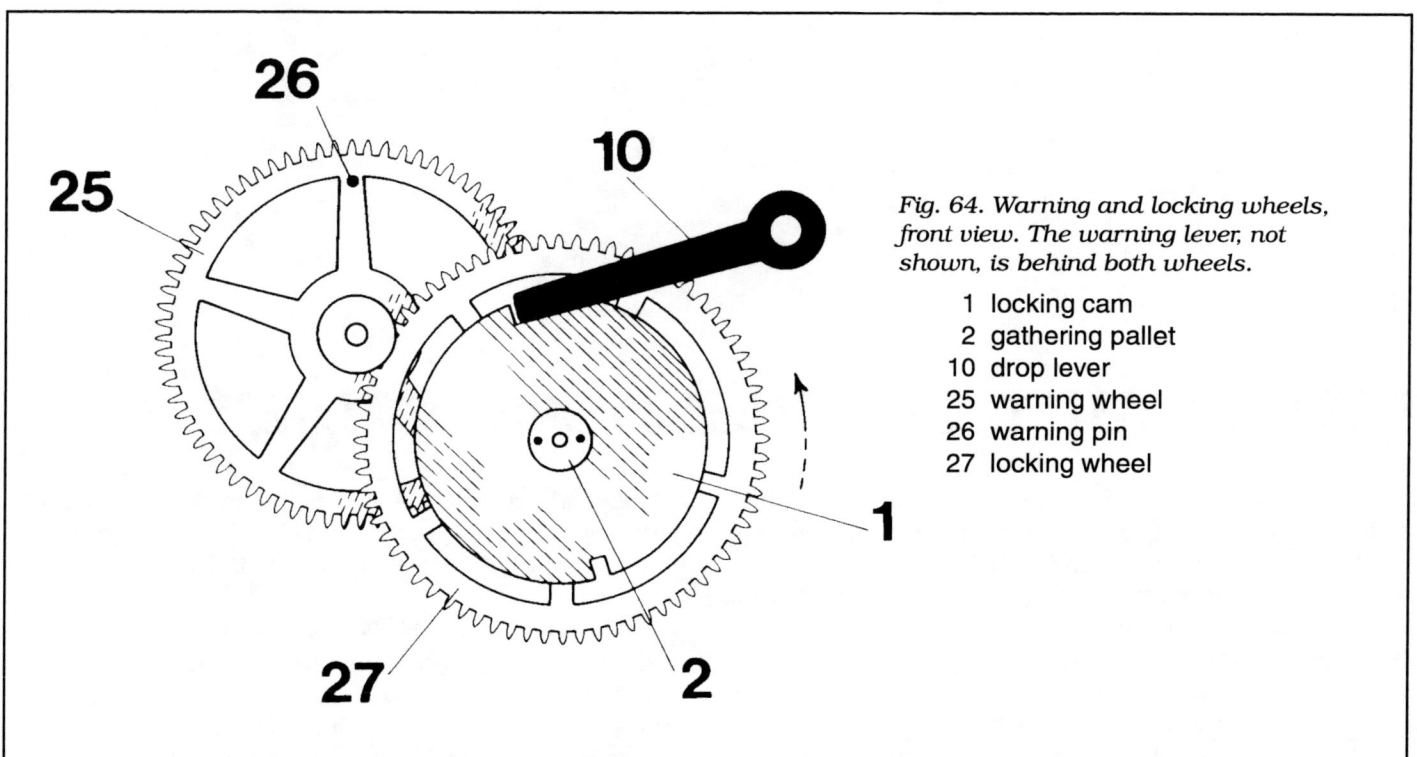

Fig. 64. Warning and locking wheels, front view. The warning lever, not shown, is behind both wheels.

1 locking cam
2 gathering pallet
10 drop lever
25 warning wheel
26 warning pin
27 locking wheel

why you may find that someone has moved the cam out of place.

To finish the assembly, install the main wheels and springs. Tighten all pillar nuts, then remove the mainspring clamps. Now put the hammer assembly back in place. Finish by adding the silencer arbor. The rest is automatic. There is no need to adjust or synchronize the pin barrel.

The worst adjustment problems I have had with this movement have had to do with the two racks. Repairers trying to apply their knowledge of normal three-train chime work to the Sessions clock often do damage. In frustration, they move the rack tails to new positions relative to the racks. Then the movement cannot count the correct number of notes. You must readjust the racks and rack tails, and check the rack springs, too. These must give positive action to the racks, or the movement will chime incorrectly.

On the quarter hours, only the quarter rack operates. It must allow up to three sequences of notes to be played. On the hour, it permits four sequences to play. The strike rack and snail determine the number of strike notes. On the hour, this is at first rather maddening to watch. The strike rack counts off first. Only when it is finished does the quarter rack come into play. This means that some or all of the chime notes are actually counted on the strike rack. All that matters is that the two racks together determine how many notes are played. One strike note equals one four-note chime sequence. The pin barrel, in conjunction with the rest of the chime mechanism, takes care of all the chime and strike counting automatically.

8

WATERBURY DOUBLE DECK CHIME

The Waterbury Double Deck movement is a manufacturer's "conversion", where a strike movement has been modified to include chime. In this instance, the chime movement is grafted onto the striker, producing a "double deck" movement with three plates. We have already covered two other conversions in earlier chapters. The Seth Thomas Sonora Chime has a separate chime movement in the case, connected by levers to a strike movement. When Sessions tackled the same problem, they added a chime mechanism onto the side of a time and strike movement.

In the instance of the Waterbury, the chime mechanism is placed on the back. Even though this creates two movements in one, they can be worked on separately. The time and strike portion is assembled first and can even be wound up before any of the chime parts are added.

Figures 65 and 66 are drawn from a movement which is probably an original or early version of the Double Deck. It has a count wheel strike and a back plate with cut-out areas. We'll use this as a basis for study. The newer model No. 35 Double Deck will be covered later in this chapter.

The front half of the Double Deck movement looks like a standard American strike movement with a count wheel. This strike movement has been modified for its purpose in a chime clock, but once it has started striking the hour, it works like any other striker. The back half of the movement is exclusively for the chime. This chime device operates on a count wheel principle, but we are used to calling this a locking plate when it is found in a chime mechanism.

The star cam on the center arbor has four arms instead of the usual one or two found in striking clocks. The cam does not start up the count wheel strike train, but rather the chime train. As the hour chime ends, the chime train puts the strike train in operation. This may sound simple enough, but one good look at the movement may convince you otherwise. There are many wheels and levers which will appear unfamiliar. Resist the urge to put the movement aside. You can tackle it step by step.

CHIME AND STRIKE OPERATION

Before thinking about disassembly and repair, follow the operation of the chime and strike mechanisms. Throughout the discussion which follows, locate the parts on Figures 65 and 66. The star cam (16) raises the chime lifting lever (17) which is connected to an arbor in the chime portion of the movement. Also on this arbor is the chime unlocking lever (6), which raises the chime drop lever (9) from the slot in the locking plate (10). This action raises the chime locking lever (4), allowing the chime lock pin (3) to escape.

The chime train moves to the warning position as the chime warning pin (2) revolves with the warning wheel toward the chime warning lever (1). There it is held until the exact quarter hour. At that moment, the arm of the star cam releases the chime lifting lever, and the warning lever drops. The warning pin is released, and the chime train begins to run. The pin barrel (8) is driven directly off the locking plate wheel to operate the hammers. The chime drop lever rides on the locking plate, and keeps the chime lock pin raised up out of the way so it will not hit the chime locking lever. The chime train runs until the drop lever reaches the next slot in the locking plate. Then the chime locking lever stops the chime lock pin and the entire gear train.

This version of the movement does not have an automatic chime correction feature. The locking plate does not have the hump or high spot found in some chime clocks. There is no chime correction cam either, which would have been made with a slot synchronized for the third quarter locking position. The four arms of the star cam are of equal length, ruling out any method of correction based on a higher lift for the hour. The chime train just counts one quarter after another. If the chime train

CHAPTER 8 - WATERBURY DOUBLE DECK CHIME

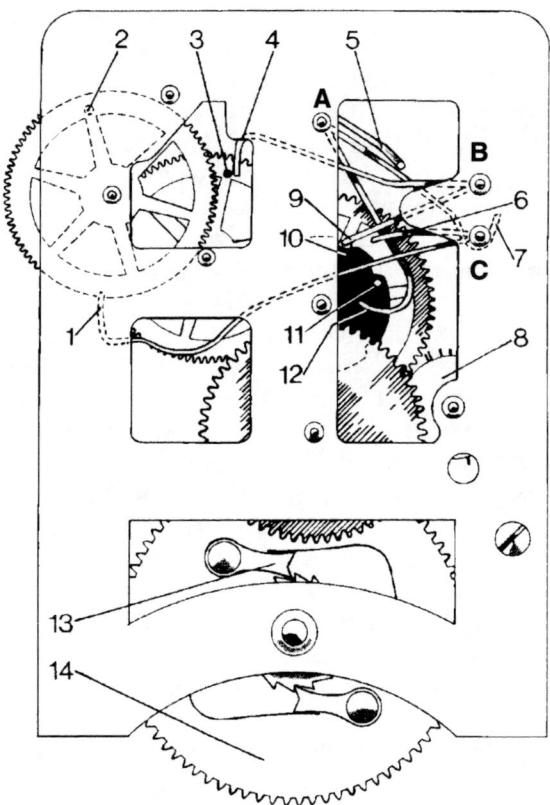

Fig. 65. Waterbury chime train, Double Deck movement, older version viewed from the rear. Arbors for the chime levers are designated A, B, and C.

1 chime warning lever
2 chime warning pin
3 chime lock pin
4 chime locking lever
5 rear strike unlocking lever
6 chime unlocking lever
7 strike warning lever
8 pin barrel
9 chime drop lever
10 locking plate
11 strike lift pin
12 strike lifting lever
13 chime click
14 chime main wheel

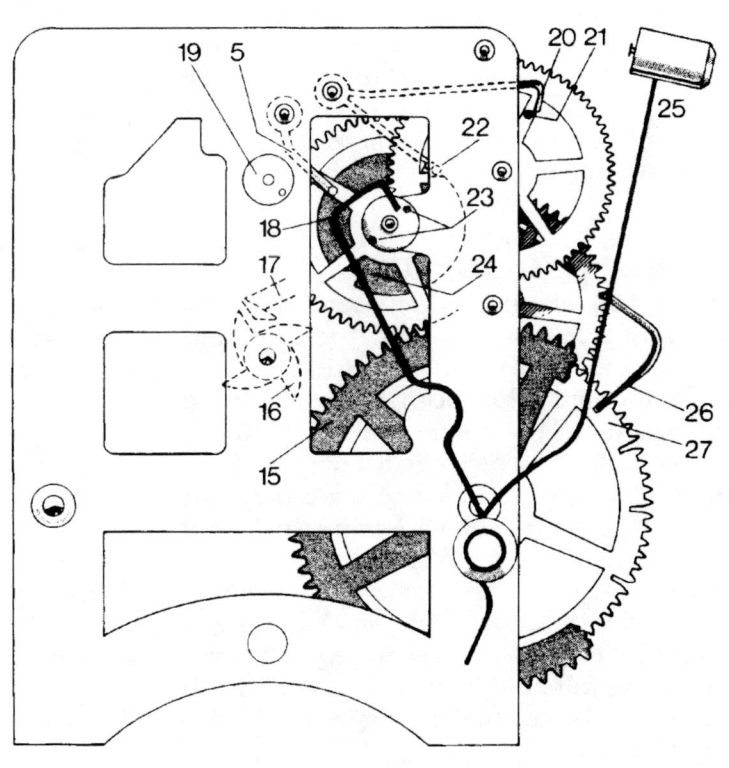

Fig. 66. Strike train (count wheel version) viewed from the rear.

5 rear strike unlocking lever
15 strike main wheel
16 star cam
17 chime lifting lever
18 strike hammer tail
19 chime silencer arbor
20 strike locking lever
21 strike lock pin
22 strike drop lever
23 strike hammer-lifting pins
24 strike cam
25 strike hammer
26 count lever
27 count wheel

runs down or someone turns the hands, the minute hand may not point in agreement with the notes being played. The only remedy is to move the minute hand counterclockwise so that it agrees with the notes played. Then the time can be set by turning the minute hand slowly clockwise, waiting for the chimes and strike to finish before going further.

The count wheel strike train does not have an automatic correction feature, nor would we expect it to have one. The movement has a trip wire hanging down, to enable the clock owner to make the clock strike ahead. In some clocks the hour hand is free enough to turn forward or back until it agrees with the hour struck, but this method invites trouble from clock owners who may loosen or bend the hand.

As the movement nears the end of the hour chime, the strike train moves to the warning position. The key to this action is the strike lift pin (11) on the locking plate. The pin raises the strike lifting lever (12). This motion also raises the strike warning lever (7), attached to the same arbor as the lifting lever. The warning lever holds the strike lock pin (21) during the warning, but it has another function to perform at the same time. The warning lever pushes up the rear strike unlocking lever (5). There are two of these strike unlocking levers. The one toward the rear of the movement is contacted by the strike warning lever as I have explained. The lever near the front of the movement (not shown on the drawings) has the trip wire hanging from it, and it is this lever which actually raises the count lever (26) up out of the slot in the count wheel. To review, the flow is from the strike lift pin to strike lifting lever, to rear strike unlocking lever, to front strike unlocking lever, to count lever. The standard count wheel striking portion of the clock takes over at this point.

DISASSEMBLY AND REPAIR

Regular mainspring clamps do not seem to fit any of the three mainsprings. However, the time and strike springs can remain wound until the chime portion of the movement has been removed. So the first step is to let down the chime spring, which is a strong one. Wrap some heavy wire around the partially wound chime spring. A single wrap may not be enough, depending on the wire you use. Use a let-down key to fully release the spring into the wire loop. On the clock studied for this chapter, the chime main wheel has double clicks. The click springs are designed to permit one of the clicks to be pried back, so the click spring will hold it away from the wheel. The other click is released in the usual way and the mainspring let down.

Before you remove the back plate, block the escape wheel. The long pallet arbor extends all the way from the front plate to the back plate, so the rear pivot will be out of its hole when you do remove the back plate. A rapidly spinning escape wheel will do damage. You may feel hesitant about taking off the back plate when the time and strike mainsprings are still wound up, so take a moment to study the way the movement is held together. The back plate is held in place with screws fitting into the movement pillars. Removing the screws and the back plate will still leave the pillars in position. They are themselves screwed into the front pillars. Picture the front part of the movement, the time and strike portion, as though it were a standard striking movement. Its back plate would be held on by nuts. In this Waterbury chime movement, the nuts have been replaced by the threaded rear pillars.

After the chime part of the movement has been removed, you must put mainspring clamps on the time and strike springs and let the springs down. The clamps will now fit, because the chime parts are out of the way. You will want to mark the main wheels, flies, warning and locking wheels for later identification. Clean the movement, install bushings, polish pivots, and take care of any other necessary work.

REASSEMBLY

The biggest concern with this movement is the reassembly procedure. Put the time and strike together first, and adjust the strike train. You can lift the strike unlocking lever manually, but the strike warning lever will not be there until the chime train is assembled. The strike warning, at least, does not have to be set or adjusted because the strike train uses the pin (21) for warning and locking. Before you leave the strike train, check for three points: 1) strike adjustments must be complete, because you cannot separate the front and middle plates again after the chime train is in place; 2) the rear pillars must be screwed on tightly, as they hold the time and strike movement together; and 3) the clamps must be removed from the time and strike mainsprings, because you will not be able to get them off when the chime portion is in place. Except for minor adjustments, we are finished with the strike train.

Before you assemble the chime train, several parts must be added. Install the chime silencer arbor (19), which is pinned to the front movement plate. The chime silencer lever, not shown in Figure 66, is screwed to the middle plate. The screw hole is located next to the front pivot hole for the chime locking wheel. Next, check the disk with the two strike hammer-lifting pins (23) fastened to it. The pins must be set so the strike hammer (25) will not

CHAPTER 8 - WATERBURY DOUBLE DECK CHIME

be left in the raised position when the strike ends. It is not possible to actually install the strike hammer at this point, so just take a trial position by placing it in its location and judging the clearance of the pins.

Install the chime main wheel next. There is also an easily forgotten part which can go in now. It is a plain arbor which acts as a hammer stop, and it fits below the pivot point for the chime and strike hammers. The second wheel, a solid wheel in the movement I studied, can go in next, followed by the locking plate wheel. The pin barrel fits in so that it is driven by the locking plate wheel. Besides the chime levers, you now have two similar-looking wheels and the fly left. The chime warning wheel can be recognized because the wheel is located nearer to the end of its arbor. The other wheel is the chime locking wheel. Do not forget to add the pallet arbor.

Before the rear plate can be installed, the three arbors which carry the chime levers must go in. This part of the job can be frustrating, because the overlapping levers shown in Figure 65 can be a real puzzle. First, locate the three sets of pivot holes for the arbors. Pick up the first arbor, C, which belongs in the lowest of the three holes. This arbor has the chime lifting lever (17, Figure 66), the chime unlocking lever (6), and the chime warning lever (1) attached to it. Install the arbor as shown in Figure 65, remembering that the chime lifting lever slips into the time and strike portion of the movement to work with the star cam.

The next arbor, A, fits in the highest of the three holes, located on the vertical center strip of the movement plate. The arbor has the strike lifting lever (12) and the strike warning lever (7) attached to it. Observe that the strike warning lever goes under the rear strike unlocking lever. If it does not, the strike train will not start, because there is no lifting action. In addition, make sure that the strike warning lever rests over the third arbor of the strike train, not under it. An incorrect installation here will mean that the warning lever will not reach the strike lock pin.

The third arbor, B, has the chime drop lever (9) and the chime locking lever (4) attached. As you install this arbor, set the chime wheels in their correct relationships. First, arrange the locking plate so that the chime drop lever falls into one of the four slots. It does not matter which slot you choose. Next, place the chime locking wheel with the chime lock pin (3) resting against the chime locking lever (4). Check the chime warning run now, by setting the chime warning wheel. When the train is in the locked position as you have set it, the chime warning wheel and pin (2) should have a half revolution run before hitting the chime warning lever. Figure 65 shows the correct setting for the wheel.

At this point, finish up the major part of the reassembly by installing the back plate and getting all the pivots into their holes. Tighten the pillar screws. Now add the chime and strike hammers. The hammers pivot on a long arbor which is threaded to fit into the middle movement plate. Add the brass spacer bushing first, then the rear chime hammer. As you turn the threaded arbor further in, install each hammer in turn. The strike hammer is closest to the middle plate, and goes in last.

The pin barrel must be synchronized with the locking plate. Loosen the set screw on the pin barrel, between the wheel and barrel. You will be able to turn the pin barrel by hand, causing the chime hammers to rise and fall without affecting the rest of the chime train. It is first necessary to establish a known chime point from which to work. Rotate the chime train, stopping after the first quarter chime. This, as we know from past repairs, is the four-note sequence of Westminster. In the movement studied, the four chime hammers work from rear to front in succession for the first quarter. Turn the pin barrel by hand, watching the hammers rise and fall. Stop when you observe the hammers move in the correct order. Tighten the set screw and then chime the movement through several quarters to make sure the hammers work properly. There is a hole in the back plate which serves as a viewing port for the pin barrel at the area where the pins lift the hammers.

NEWER VERSION MANSION NO. 35 MOVEMENT

There is another, probably more recent version of the Waterbury chime movement, which has some important design differences. Figure 67 is a photo of the No. 35 movement from the "Mansion" series of chime clocks. The picture reveals several basic differences when it is compared to the "earlier" movement we have studied in this chapter. The movement in the catalog illustration was "made with solid back plate; rack strike; reduced wind feature . . ." In the example I studied, the back plate was stamped with eight patent dates extending from 1910 to 1925.

I have already mentioned the chime winding setup. It is difficult to wind the powerful chime mainspring in the older movement. For the newer version, Waterbury added a "reduced wind feature" to gear down the winding effort. It means more turns to wind the spring, but the turning is easier. A going barrel replaces the exposed loop-end spring. Fortunately, the double chime clicks were retained.

The older movement did not have a chime correction mechanism. If the sequences of notes were

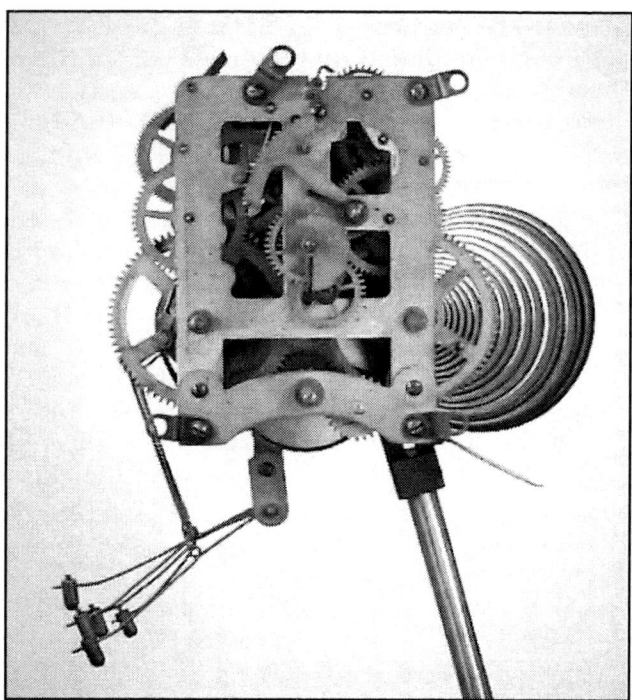

Fig. 67. No. 35 Double Deck chime movement (newer version) from the Mansion line by Waterbury. This movement was loaned by Bill Rakes. Note the broken time mainspring on the right side.

thrown off for any reason, they did not self-correct. Although the hour strike would always follow the hour chime, the chime could come at any quarter if upset. A chime correction mechanism was added for the newer version. The strike mechanism in the old movement was the count wheel type. This is not, of course, a self-correcting device. Each hour counts in succession, even if an error occurs because of hand movement or lack of winding. Especially in a chime clock, this system leaves a lot to be desired. Chime and strike can both be wrong. A clock equipped this way can chime and strike 10 o'clock at 4:30, for example. The change-over to rack and snail in the newer version made for a more reliable clock. Chime and strike will automatically reset themselves as long as the hands are installed the right way.

Moving Things Around
For these changes to be made, the chime and strike levers had to be rearranged. Figures 68 and 69 show the chime train layout, plus the rack and snail strike, in the "new" No. 35 chime movement. The addition of automatic chime correction caused the most change in the movement. Figure 68 is a diagram showing the locations of the major chime parts in the newer version. Waterbury made the changes without adding any new arbors between the plates in the chime end.

Consider the three arbors which carry the chime levers. One is located high up, near the centerline of the back plate. Two others are on the right edge, one above the other. The arbors are marked A, B, and C. The same three arbors are found in the old movement. Arbor A still carries the strike warning lever (7), although the shape is different. The strike lifting lever (12) was moved entirely out, to the strike portion of the movement (see Figure 69). The chime drop lever (9) is now located on arbor A. Chime locking was moved one wheel higher up in the train. A chime locking cam (28) replaces the older pin. A redesigned chime locking lever (4) is fastened to arbor A.

Arbor B now has the chime correction lever (30) and chime correction arm (31). In the old movement, the chime locking lever (4) and chime drop lever (9) were located here. They had to be moved to arbor A, as mentioned.

Arbor C carries the same three levers as before. The chime warning lever (1) is unchanged. Note that chime warning and locking are now on the same wheel. The chime unlocking lever (6) is still there, but reshaped. And the chime lifting lever (17 on Figure 66, but not shown on Figure 68), is the same. The star cam (16, shown only in Figure 66) raises the chime lifting lever each quarter hour to start the chime. This is just plain confusing, but diagrams of these levers can save you a lot of time during a repair, when you really need the information.

Chime Correction Feature
Taken by itself, the chime correction mechanism is fairly simple and nothing new. Seth Thomas used a similar device in the No. 113 chime movement (Chapter 4). Arbor B in the Waterbury has a three-fingered blue steel spring (like the kind for hand tension) to stiffen its action; the arbor stays where it is moved, until it is moved again. This is the key to the chime correction mechanism. The locking plate (10) has a pin on it which pushes the chime correction arm (31) as the third quarter chime ends. This moves the chime correction lever (30) down into the path of the chime correction pin (29). The dotted lines show the lever in position. The chime train will unlock at the next quarter, but unless the minute hand is pointing straight up, the mechanism will not run. The lever (30) stops the pin (29). Unlike the old movement, the newer version has a star cam with one arm longer than the other three. The longer one gives enough lifting action to raise the chime correction lever up and out of the way. And since the tension spring holds the lever, it stays up. At the end of the third quarter chime, the process starts all over again.

There are two basic adjustments to consider. One involves the assembly of the wheels. The wheel car-

CHAPTER 8 - WATERBURY DOUBLE DECK CHIME

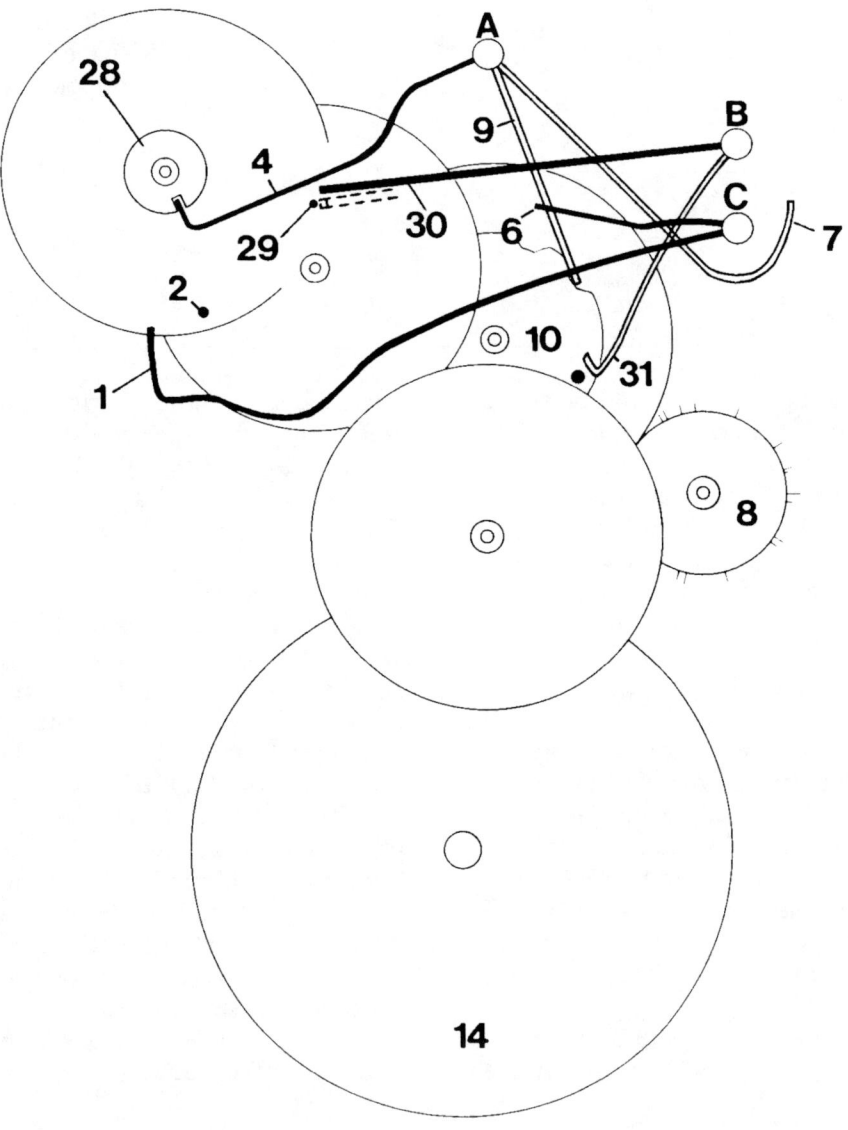

Fig. 68. Waterbury chime train, newer version, viewed from the rear. Arbors for the chime levers are designated A, B, and C.

1 chime warning lever
2 chime warning pin
4 chime locking lever
6 chime unlocking lever
7 strike warning lever
8 pin barrel
9 chime drop lever
10 locking plate
14 chime main wheel (barrel)
28 chime locking cam
29 chime correction pin
30 chime correction lever
31 chime correction arm

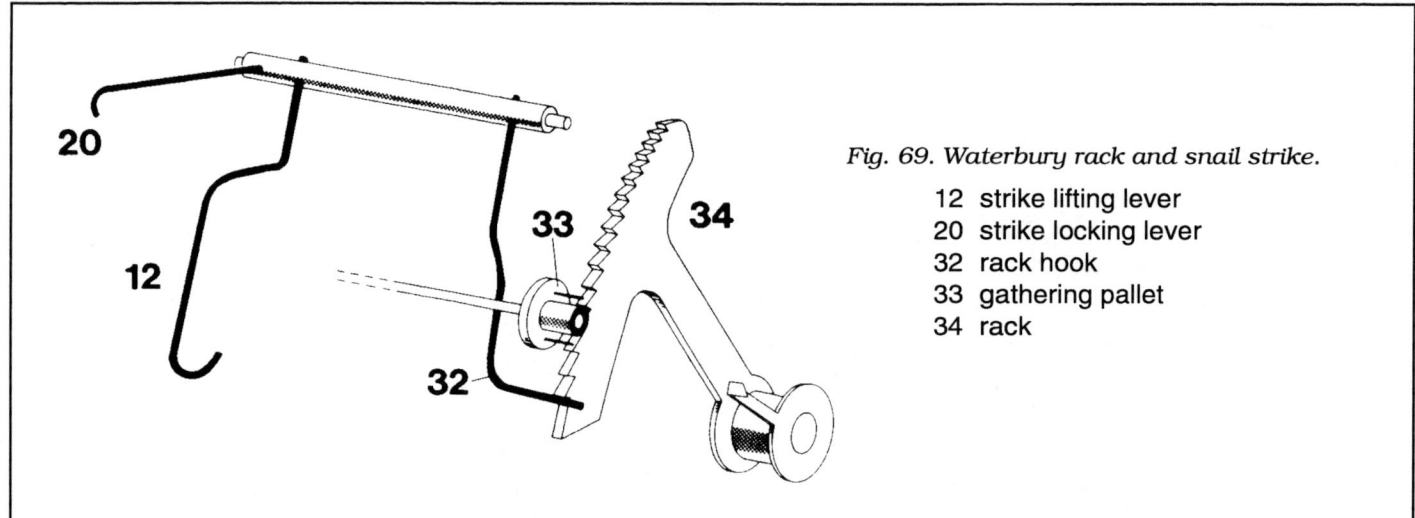

Fig. 69. Waterbury rack and snail strike.

12 strike lifting lever
20 strike locking lever
32 rack hook
33 gathering pallet
34 rack

rying the chime correction pin (29) should be set to place the pin close to the chime correction lever (30). If it is too far away, the chime will begin before the pin and lever come together. The other adjustment is in the chime correction lever. Bend it if necessary to insure that it stops the pin during the correction phase and clears it as the chime train runs.

Rack & Snail Strike

The strike mechanism is actually simpler than the older count wheel. There is only one arbor to carry three levers. Figure 69 shows the parts. The strike lifting lever (12) is in this portion, having been moved out of the chime portion of the old movement. A pin on the locking plate raises the lever as the hour chime ends, to start the strike train. Another lever in this area is the strike locking lever (20). This is in the same location as in the old movement. The lever can be bent up or down to adjust the locking action.

The rack hook (32), gathering pallet (33), and rack (34) are also shown in Figure 69. Always remove the gathering pallet during repair, because you will need to adjust it carefully as you reassemble the strike train. The pallet has two steel pins, which should be clear of the rack teeth as striking ends. This is essential, to permit the rack to fall at the next hour. The pin wheel also has two pins. Position the wheel so that the pins are free of the hammer tail when the clock isn't striking. The strike warning action, at least, is "automatic". You don't have to set or adjust it, because the same pin serves for both locking and warning. The pin and wheel just run from one lever to the other as the train changes from lock to warning.

The Waterbury "Double Deck" movement is a challenge, to say the least. It is hard to retain your "experience factor" on it between jobs. It isn't as common as other chime clocks, and there is much here that is particular to this movement alone. Add to this the fact that there are at least two rather different versions of the movement, and you have an even greater challenge. The illustrations in this chapter will help.

9
ANSONIA

This chapter features an American chime movement by Ansonia. One example I studied was housed in a large tambour case 22-1/2 inches wide. This is typical of the style popular during the 1920's and '30's: wide cases with heavy movements, designed for the mantelpiece. The material which follows is based on the restoration of an Ansonia chime clock made in 1926. The movement was not in running condition: gummy oil and dirt were the main problems. After taking down the movement, I saw that most of the pivot holes were in good shape, although the pivots needed polishing. The three mainsprings were found to be usable after cleaning.

Overall dimensions of the Ansonia chime movement are 4-5/8 inches wide by 5-3/4 inches high. The three mainsprings are contained in barrels. Each winding mechanism has an extra wheel to gear down the ratio and ease the effort of winding the powerful springs. Although some movements of this type have double clicks at least on the chime, the Ansonia does not. Failure of the clicks or click springs brings serious damage, usually in the form of broken barrel teeth. The greatest service problem in older spring driven chime movements is the potential for this kind of destruction.

We begin by studying the front movement parts. Then we go behind the front plate for more details of the chime and strike, including the chime self-correction mechanism. We will finish up the Ansonia with a study of the hammer assemblies and silencer, and some hints on repair.

CHIME
The chime train is the locking plate type. A look at the front of the movement in Figure 70 shows a cam on the right side, where the locking plate is usually located. This part is actually the chime correction cam, not the locking plate. Notice that it has only one slot instead of the four for the locking plate. Figure 71 shows the locking plate (26), located between the movement plates. Once this confusion is cleared up, the Ansonia movement is easier to understand.

Chiming begins as the star cam (not shown) raises the lifting piece (11). This piece acts upon the chime unlocking lever (6). The chime and strike levers overlap each other, making it difficult to trace them. The chime unlocking lever raises the strike lifting lever (16), causing several things to happen at once. We'll trace each one.

The strike lifting lever moves the strike warning lever (5), affecting the strike train. We will cover this action later under "STRIKE". Behind the front plate, three other levers are attached to the arbor which supports the lever (16). These are the chime locking lever (21), the chime drop lever (24), and the chime lock piece (25). As we have seen in other movements, the chime locking lever (21) stops the chime lock pin (20) to hold the chime train. When the lever (16) is raised, the locking lever releases the pin, allowing the chime locking wheel (19) to turn. At the same time, the chime lock piece comes up from a slot in the locking plate to determine the length of each chime sequence.

The other action occurring at the same moment involves the chime drop lever (24). As it is raised out of the slot in the cam (23), the chime train moves toward the warning position. This action is enough to move the slot out from under the drop lever. Then, at the moment chiming begins, the drop lever falls to the rim of the cam. The higher position (compared to the slot) keeps the chime lock piece from dropping back into the slot in the locking plate. If it did drop back, chiming would not happen at all. Remember that the four levers (16), (21), (24), and (25) are all fastened to the same arbor and move together.

The chime warning lever (22) stops the chime train at warning. When the chime locking lever (21) releases the pin (20), the warning lever is already in position. It stops the pin and holds it until chiming begins. The chime continues until the chime lock

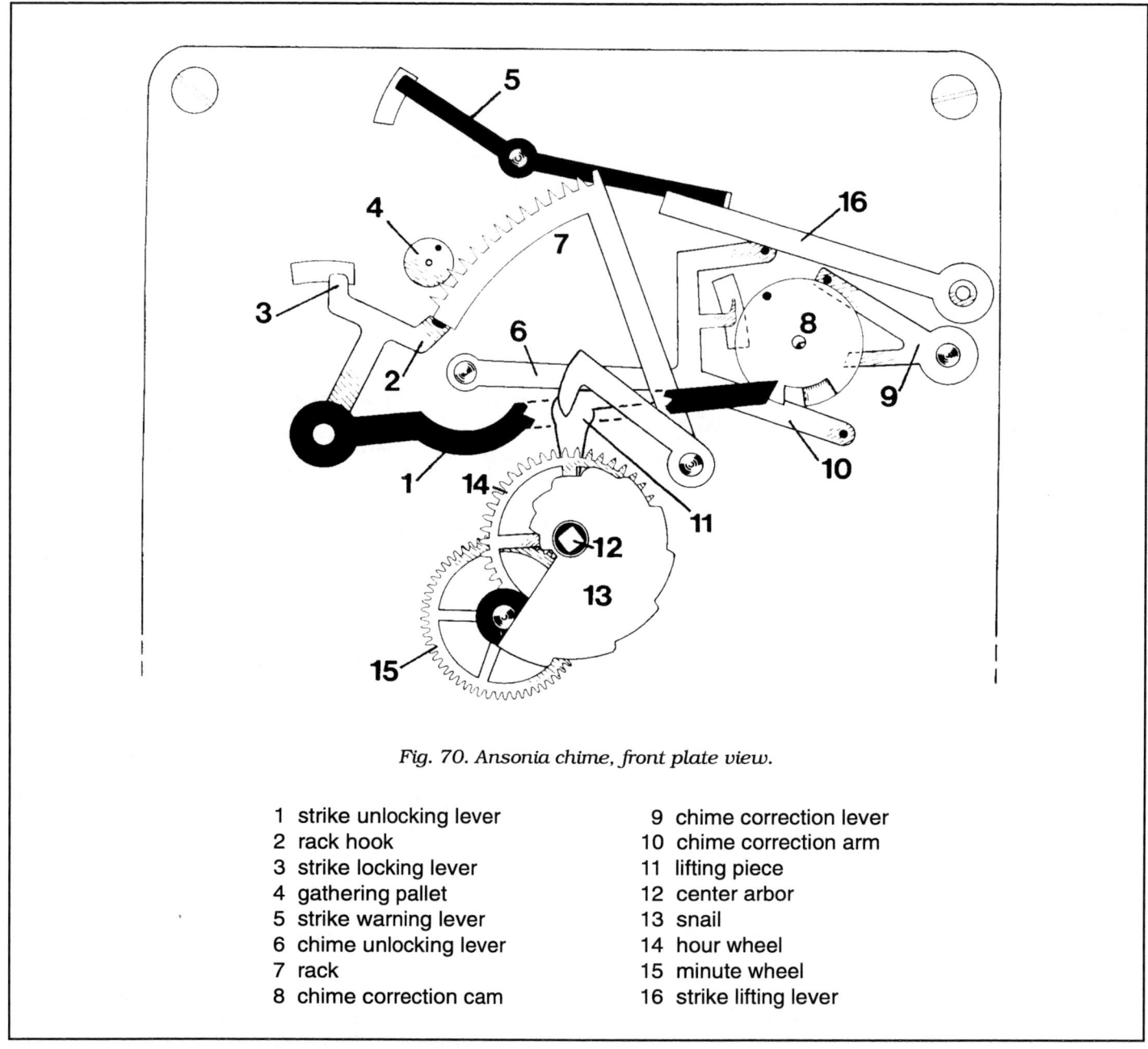

Fig. 70. Ansonia chime, front plate view.

1 strike unlocking lever
2 rack hook
3 strike locking lever
4 gathering pallet
5 strike warning lever
6 chime unlocking lever
7 rack
8 chime correction cam
9 chime correction lever
10 chime correction arm
11 lifting piece
12 center arbor
13 snail
14 hour wheel
15 minute wheel
16 strike lifting lever

piece drops into the next slot in the locking plate. This action places the chime locking lever in the path of the chime lock pin.

Chime Correction

The automatic chime correction feature is a mechanism which appears in many different forms in chime clocks. At least the Ansonia mechanism is easy to understand. The fact that it is on the front of the movement, instead of buried inside, makes it easier to observe. Like most of these devices, it corrects the chime sequences by preventing the hour chime from sounding except on the hour. And since the strike is started up by the chime, it is automatically going to fall upon the hour as well. Figures 72, 73, and 74 show how the chime correction mechanism works.

Begin with Figure 72. The chime correction lever (9) rides on the chime correction cam (8) as the clock chimes. The cam holds the lever high, keeping the chime correction pin (29) clear of the chime correction lever. Chiming proceeds without any interference.

Figure 73 shows the mechanism in the chime correction mode, blocking the chime correction pin. Here's what has happened: at the end of the third quarter chime melody, the chime correction lever dropped into the single slot in the chime correction cam (8). At the next chiming point, which was not the hour, the chime gears went to the warning po-

CHAPTER 9 - ANSONIA

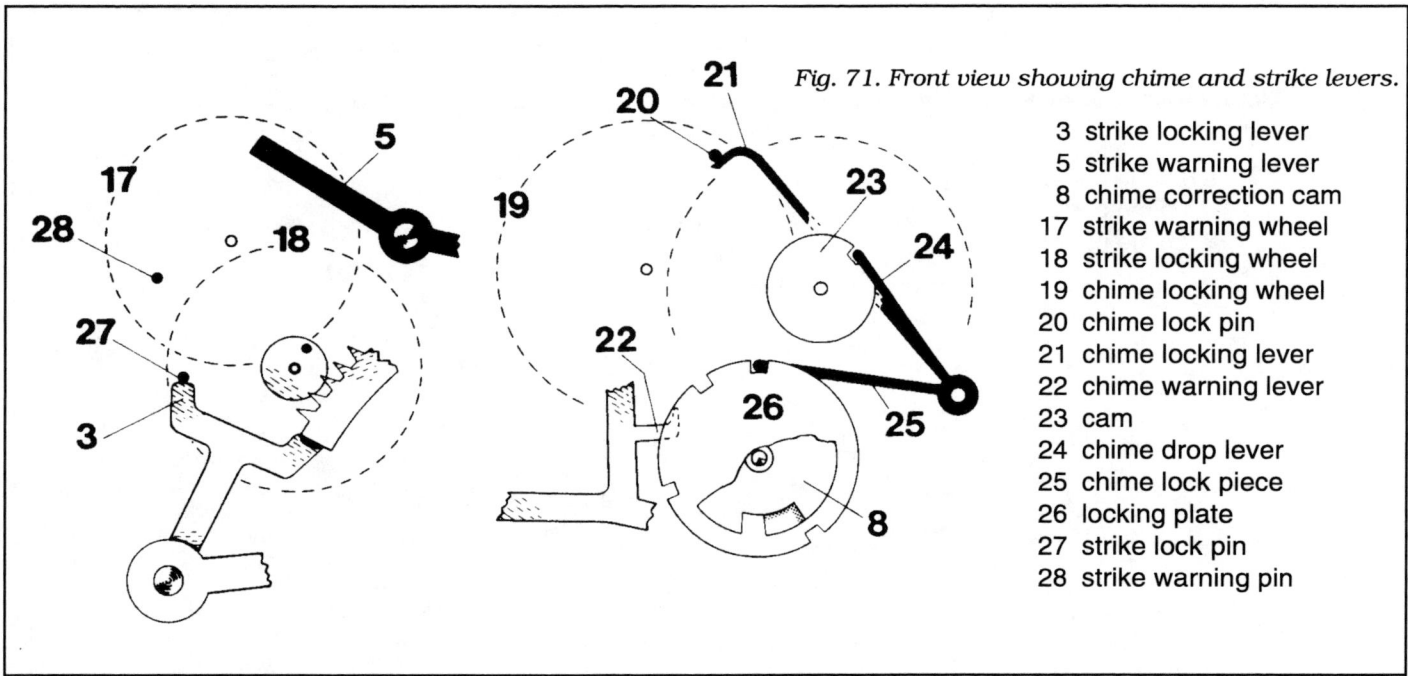

Fig. 71. Front view showing chime and strike levers.

- 3 strike locking lever
- 5 strike warning lever
- 8 chime correction cam
- 17 strike warning wheel
- 18 strike locking wheel
- 19 chime locking wheel
- 20 chime lock pin
- 21 chime locking lever
- 22 chime warning lever
- 23 cam
- 24 chime drop lever
- 25 chime lock piece
- 26 locking plate
- 27 strike lock pin
- 28 strike warning pin

sition and then began to turn. Almost immediately, before a single hammer was raised, the chime correction lever stopped the pin. That is the instant of time shown in Figure 73. A maximum of three chiming opportunities are blocked until the chimes are synchronized. The only time the pin touches the lever (9) is in the case of chime correction. Otherwise, they do no t meet.

Figure 74 indicates how the chime is unlocked for the hour. The chime correction arm (10) is raised each quarter hour, but not high enough to unlock the mechanism. At the hour, the correction arm lifts high enough to raise the lever (9) out of the slot in the chime correction cam (8). The cam slot contains a spring-loaded gate. As the lever comes up, the gate slips under the pin. In this way, the chime correction feature is cancelled out.

STRIKE

The strike mechanism is the rack and snail type. Refer to Figures 70 and 71. The gathering pallet (4) is a disk with a pin for gathering the rack teeth. The rack hook (2) and the strike locking lever (3) are part of the same y-shaped piece. To start the strike, a pin on the chime correction cam raises the strike unlocking lever (1) during the hour chime. The unlocking lever is fastened to the y-shaped piece which makes up the rack hook and locking lever, so they all move together. As the locking lever moves, the strike lock pin (27) is released. The strike warning pin (28) moves around to the strike warning lever (5).

At this point it is necessary to explain the operation of the strike warning lever. Before the hour chime even begins, the strike lifting lever (16) pushes up the right side of the strike warning lever. This pivots the left side of the lever downward, where it will be in a position to stop the warning pin.

Just before the end of the hour chime, the pin on the chime correction cam releases the strike unlocking lever. The rack hook then seats itself in the rack. The location depends upon the rack tail, which falls to a step on the snail (13). As the hour chime concludes, the strike lifting lever (16) drops, removing pressure from the strike warning lever (5). The warning lever is spring-loaded in the "run" position, so it readily releases the warning pin. The strike train proceeds to run until the gathering pallet has counted off the last rack tooth. As the rack hook slips under the end of the rack, the strike locking lever stops the lock pin.

HAMMER OPERATION

The chime hammer assembly is powered by the chime drive wheel (40), shown in Figure 76. The wheel meshes with another on the end of the pin barrel (43). In turn, the pin barrel moves the upper hammer assembly, which is hooked up to the lower assembly by the hammer chains (44).

After the chime train has been reassembled, the hammer sequences must be adjusted. In most chime clocks, it is best to loosen the set screw on the chime drive wheel, then turn the hammer assembly by itself, to a known point. The first quarter note sequence is easiest to identify. For this reason, operate the chime train to the end of the first quarter chime, then move the hammers to agree. In this clock, the four descending notes are the result of

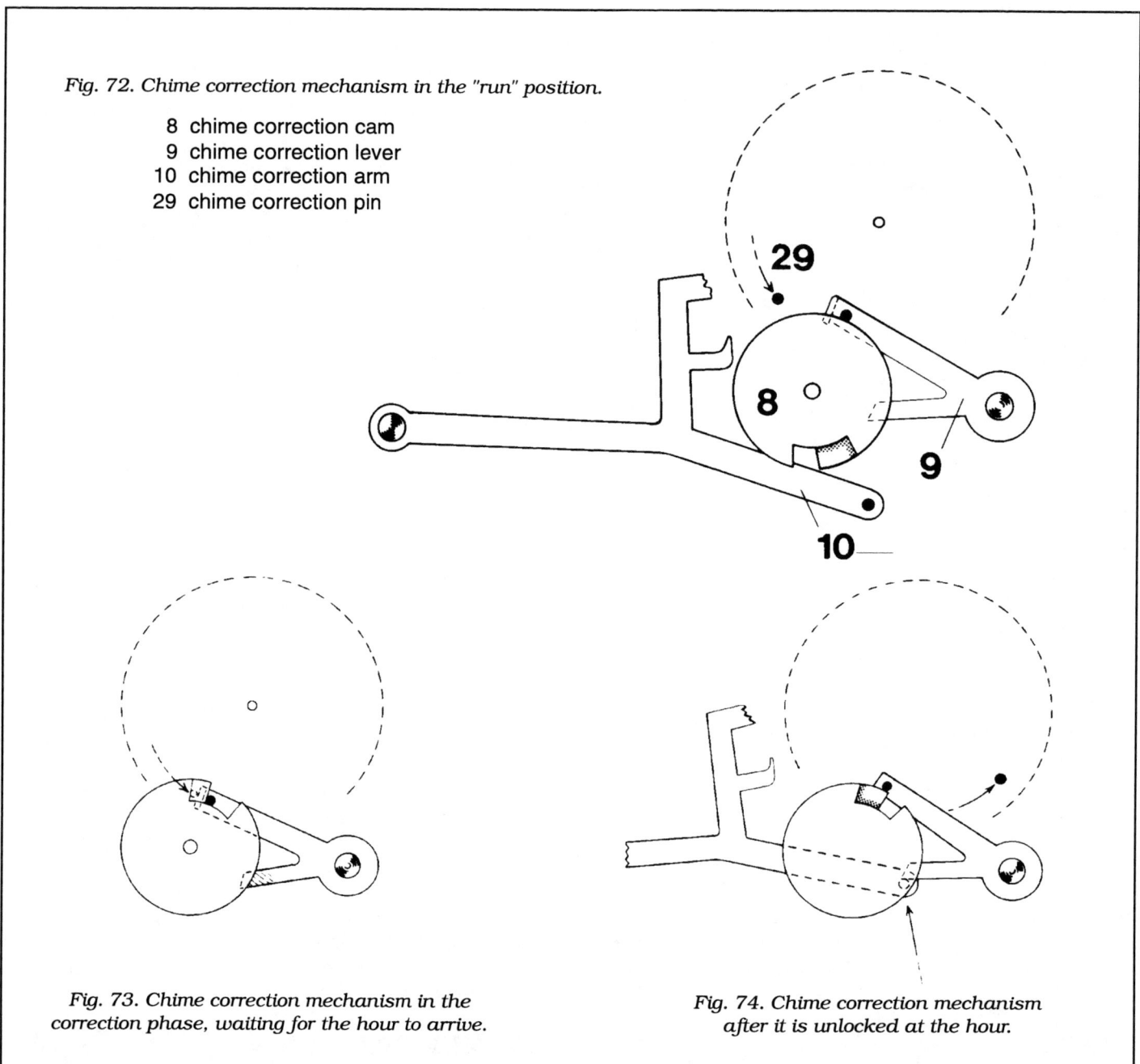

Fig. 72. Chime correction mechanism in the "run" position.

    8 chime correction cam
    9 chime correction lever
  10 chime correction arm
  29 chime correction pin

Fig. 73. Chime correction mechanism in the correction phase, waiting for the hour to arrive.

Fig. 74. Chime correction mechanism after it is unlocked at the hour.

the chime hammers hitting in order from back to front.

Unfortunately, the set screw on the chime drive wheel is small and hard to reach on the Ansonia chime. It is easier to tighten the screw as you put the movement together, before you install the chime assembly. When you want to adjust the chime sequences, rotate the pin barrel and reach the desired note pattern first. Then, without changing anything, install the hammer assembly. The pin barrel gear and the chime drive wheel should mesh correctly.

Strike hammer action begins with the hammer-lifting star (48), which raises the hammer tail (49). The strike hammer-lifting arm (42) carries the motion across the back of the movement, to the strike hammer. The arm is next to the rear movement plate, and it pivots with a seesaw motion. The strike lift pin (50) pushes up, which moves the other end of the arm downward. This action raises the single strike hammer. The hammer falls as the hammer-lifting star turns, releasing the pin (50).

The chime silencer (47) is a lever which can raise the four chime hammers to silence them. The silencer arbor pierces the dial and runs through the movement. A coil spring, pin, and washer arrangement presses the arbor against the rear plate. This creates enough resistance to keep the lever wher-

CHAPTER 9 - ANSONIA

ever it is turned. By inserting the smaller end of the winding key and turning the silencer arbor, the owner moves the silencer lever into contact with the hammer assembly. There is no silencing effect on the strike hammer. You can leave the silencer arbor installed in the rear plate when you take apart the movement.

NOTES ON REPAIRING THE MOVEMENT

Figure 76 shows the rear of the movement. The upper hammer assembly (41) and lower hammer assembly (46) are held together by the chime bracket plate (45). The hammers extend well below the bottom of the movement, but this isn't as awkward as it seems. Remove the plate screw just below the pin barrel (43), and turn the bracket at an angle. This will make it easier to get the movement out of the case.

Once the movement is out, you can take off the upper and lower hammer assemblies along with the pin barrel, as a single unit. Then decide whether to take the assembly apart. If it moves freely after ultrasonic cleaning, you may want to leave it together. With most of the older chime movements you don't have this choice, because you must remove the hammer levers and the spacer washers between them to take the movement itself apart. Check the condition of the hammer chains (44), and the five hammer heads. Repair as necessary.

Figure 75 shows the lower front movement plate (30). Do not loosen or remove any winding parts until all three mainsprings are fully let down. The lower front movement plate allows you to remove the barrels without taking the movement completely apart. In most instances, the movement needs to come apart anyway. But at least the repair is made easier by the fact that you can take out the barrels at the beginning of the job, and keep them out of the way until the work is nearly complete.

The lower front movement plate supports the three winding mechanisms. The time mechanism (39) is in the center. The cover plate (33) is cut partly away to expose the click wheel (34) and winding wheel (35). With the power safely off, strip all the winding parts from the lower movement plate. Leave the click (36) and click spring (37) in place unless they need repair. External winding parts from time, strike, and chime are interchangeable.

After removing the lower plate, take out the barrels. Mark or identify each barrel, cover, barrel arbor, and spring so you don't get them mixed up during cleaning. A good mainspring winder is essential for the job of safely getting the springs out of the barrels and then back in after cleaning. You cannot expect a successful repair unless the springs are clean. Check for cracked ends or a "set" condi-

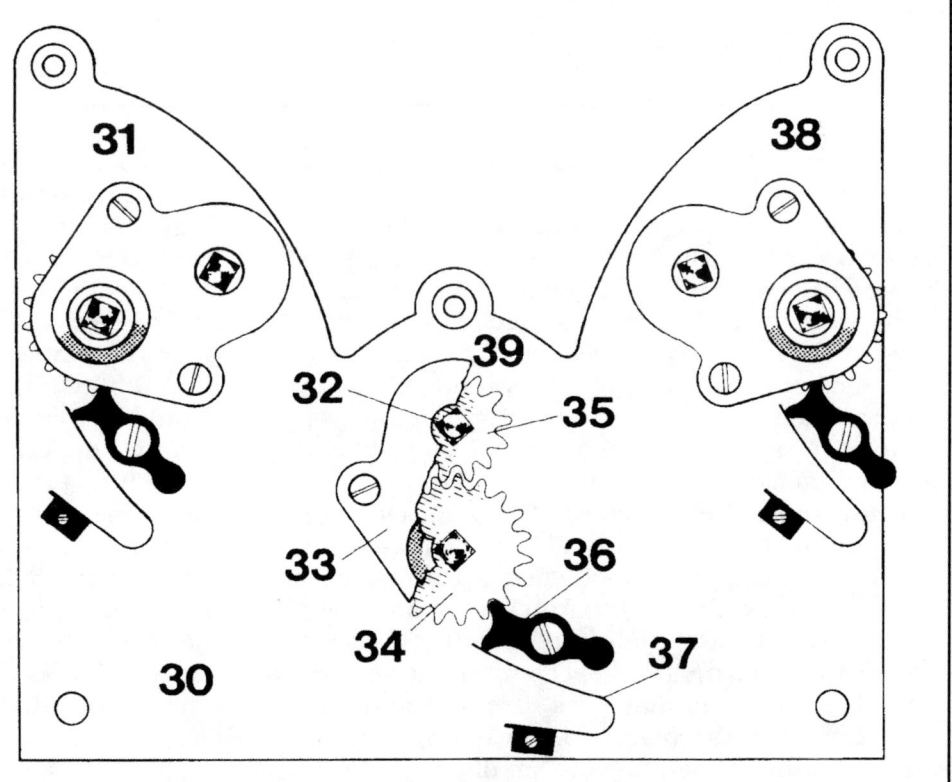

Fig. 75. Lower front movement plate, with cut-away view of the time winding mechanism.

30 lower front movement plate
31 strike winding mechanism
32 winding arbor
33 cover plate
34 click wheel
35 winding wheel
36 click
37 click spring
38 chime winding mechanism
39 time winding mechanism

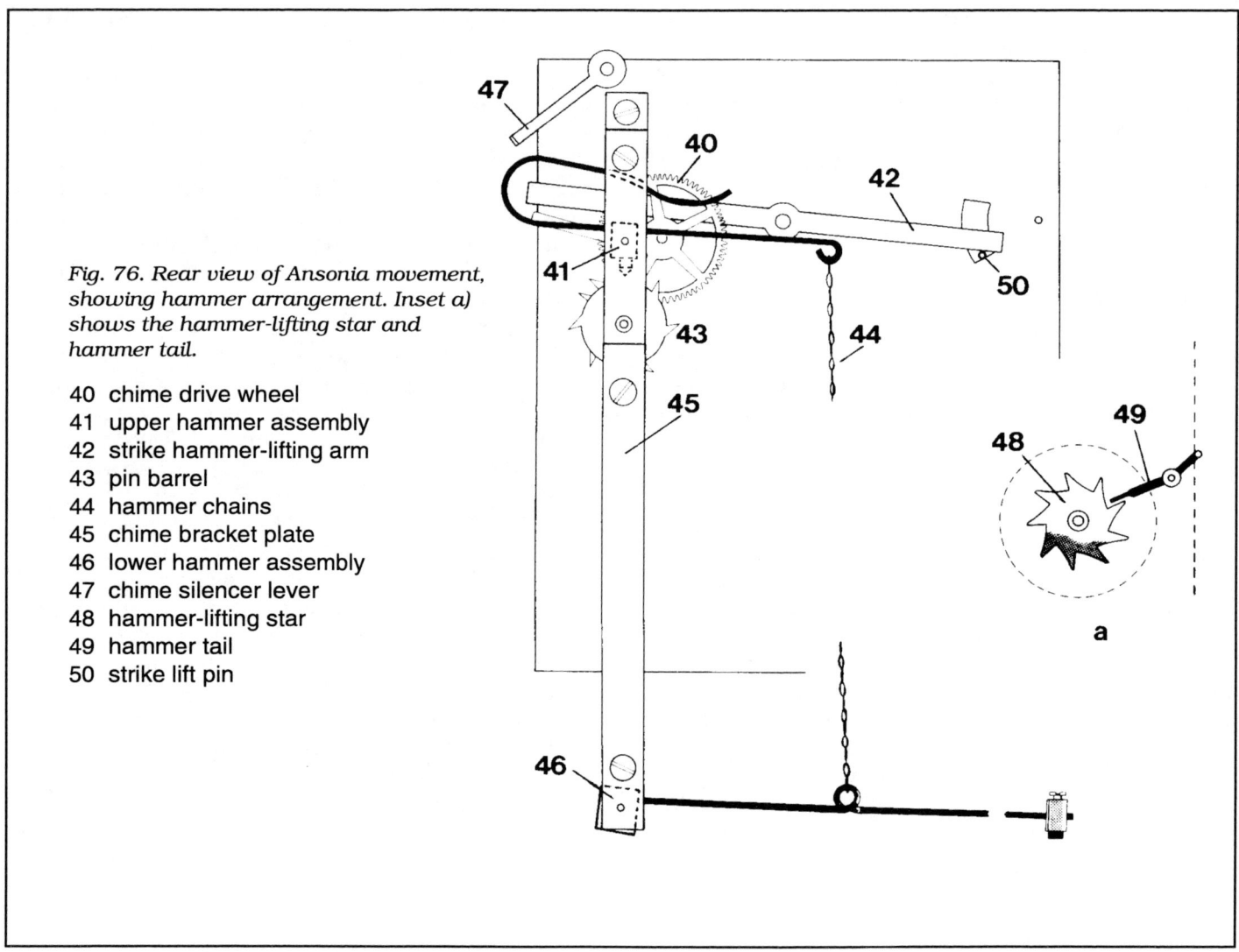

Fig. 76. Rear view of Ansonia movement, showing hammer arrangement. Inset a) shows the hammer-lifting star and hammer tail.

40 chime drive wheel
41 upper hammer assembly
42 strike hammer-lifting arm
43 pin barrel
44 hammer chains
45 chime bracket plate
46 lower hammer assembly
47 chime silencer lever
48 hammer-lifting star
49 hammer tail
50 strike lift pin

tion. It should be obvious that a bad mainspring will cause the clock to fail.

Inspect the winding wheels and click wheels for excessive wear or bad teeth. If exact replacements for these are not available from suppliers, it might be necessary to make or order a specially cut wheel. Wear is often the worst on the chime winding parts, because of the stronger spring. As the teeth wear, winding action becomes bumpy and stiff. If the click fails to seat itself as you turn the key, you will end up with damage.

To assemble the movement following cleaning and repair, install all the arbors into the rear plate. Don't forget the arbor which carries the hammer tail (49) and strike lift pin (50). You cannot add these arbors later, after the plates are together. Add the upper front movement plate, easing the lower pivots in place first. You may want to leave the two fly arbors out until the plate is halfway on, to minimize the chance of bending the small pivots. Never force them, and never forget they are there. The re-

coil pallets should be added along with the fly arbors. You can still add the pallets at the end if you forget to include them. Just separate the plates enough to allow the pivots to go into the holes.

Strike Adjustments

After the plates are together, install the strike levers as illustrated. The strike hammer-lifting arm should be free to move without sticking or binding: hammer action will be sluggish unless there is a minimum of friction. Turn the wheels by hand and check the strike operation. As the strike finishes, the hammer tail must be free of the hammer-lifting star. Figure 76 (inset a) shows the correct position. If necessary, separate the plates enough to adjust the mesh between the third wheel and the pinion above it. Look at the position of the pin on the gathering pallet at the same time, to make sure it is clear of the rack. Remove and reinstall the gathering pallet if necessary, to correct it.

The strike warning pin (28, Figure 71) should

CHAPTER 9 - ANSONIA

have a 7 or 8 o'clock rest position. It can then travel almost a half revolution of the wheel as warning occurs. The only other strike adjustment is the snail position for 12 o'clock. This is probably one of the last items to take care of during assembly of the movement. It is easier to take care of after chime adjustments are completed.

Chime Adjustments

Refer to Figure 71 for parts identification. The chime correction cam (8) should be added to the movement. Turn the wheels around to the end of the third quarter chime. With the chime lock piece (25) seated in the third quarter slot in the locking plate (26), the chime correction cam (8) should be tightened in place as in Figure 73.

Next, observe the chime drop lever (24). It should rest in the slot in the cam (23). In addition, the chime lock pin (20) should rest against the chime locking lever (21) to stop the chime train. You may have to separate the plates and adjust these parts several times to get them right. The design is similar to that of the Seth Thomas No. 124 and 113 movements. Chime warning does not need to be set up; the chime lock pin moves from the locking lever to the warning lever.

Rotate the chime locking wheel (19) to seat the chime lock pin (20) against the chime locking lever (21). The chime correction pin (29) should now be oriented at about a 7 o'clock position. This will allow the chime correction pin to come into play during the correction phase.

To finish up, add the mainspring barrels and the lower front movement plate. Install the winding parts shown in Figure 75. After all the pillar screws are tight, you are ready to wind and test. The real trial of the movement is a solid eight-day run. You will have to wind the springs all the way up to make it. Wind carefully, because the geared-down winding mechanisms take away much of the "feel" of winding. It is very hard to tell when the springs are almost wound up. Make sure your customer knows how to wind fully without forcing the key.

Mainsprings

The mainsprings were removed from an Ansonia movement so that width and length could be measured. Rather than try to stretch the mainsprings out for a length measurement, I calculated the ideal length using the formula below.

Mainspring Length in Five Steps
1. Inside diameter of barrel, squared, times .7854
2. Diameter of arbor, squared, times .7854
3. Subtract step #2 from step #1
4. Divide by 2
5. Divide by the mainspring thickness

The chime mainspring for the Ansonia chime movement is 1" wide x .018" thick. I calculated the ideal mainspring length for the barrel as 69". A supply catalog listed a 96" long spring which could be shortened to 69" as a replacement.

The time mainspring is 7/8" wide x .017" thick. The length was calculated as 73". The closest spring I found in a catalog was 7/8" wide x .0175" thick x 70" long.

The strike mainspring is 7/8" wide x .016" thick. The ideal length is 77-1/2". Catalog replacements did not closely match this spring. Perhaps the spring mentioned above as a time spring replacement would be the closest selection.

10

GLOBE

The Globe chime movement is included here as a reference for anyone who may have the opportunity to repair this relatively rare movement. It is a side issue in chime clock repair, certainly not one of the mainstream movements we see often. Perhaps the most interesting feature of the Globe is that it has only two gear trains, one for time and the other for chime and strike. The usual chime clock arrangement is, of course, a three train setup with time, chime, and strike geared separately.

The front movement view is shown in Figure 77. The rear view in Figure 78 omits certain parts such as the pin barrel, which would have hidden other parts more important for our study. The clock case for this example is an ordinary tambour 21 inches wide and 10-1/2 inches high. The dial is marked "Made by Globe Clock Co., U.S.A.". The movement has heavy plates, all solid pinions, and barreled mainsprings. It is a five hammer Westminster chime model, with the fifth hammer for the strike.

With one train for chime and strike, some provision must be made for switching from chime to strike at the hour. Some of the clock manufacturers who have come up with workable mechanisms include Seth Thomas (Sonora), Sessions, and Herschede. A movement by the Hamburg-American Clock Company was the type J.E. Coleman called a "cheater". It operated on two trains but played Westminster chimes on the first three quarters: the hour had the strike but no chime. The Globe movement featured in this chapter is full Westminster chime.

The Globe has its own unique way of making the change from chime to strike at the hour. In trying to understand the workings of any two train chime clock, this is the first thing that should interest the clockmaker. As the clock chimes, the chime drive pinion (18) turns the pin barrel wheel (23) and pin barrel (not shown) to lift the chime hammers (13). The pin barrel wheel has two teeth cut out of its edge. Note the sliding gear section (22) with two teeth attached to it. When the sliding gear section is oriented with the teeth pointing up, the section slips down to make a break in the wheel. The open segment of the wheel reaches the chime drive pinion just at the end of the 16-note hour chime. The pinion continues to turn with the wheel disengaged, and striking begins.

STRIKE

Striking is accomplished by the hammer-lifting star (21) on the third wheel of the chime-strike train, acting upon the strike hammer tail (19). The strike hammer linkage (20) raises the strike hammer. When the clock is chiming, the hammer tail is held away from the hammer-lifting star to prevent striking. The strike silencing lever (17) rides on the strike cam (15) and as it does so, it holds up the strike hammer tail. There is one slot in the cam, which comes by just at the end of the hour chime. When the strike silencing lever drops into the slot, the hammer tail is lowered so the hammer-lifting star can act upon it to raise the strike hammer linkage.

At the first quarter, the pin barrel wheel must be reengaged with the chime drive pinion so the chimes will work again. By about ten minutes past the hour, the sliding gear lever (16) has lifted the sliding gear section back up, completing the circumference of the pin barrel wheel. Exactly at the first quarter, the sliding gear lever drops, but the gear section is held in place by the support piece (24) until the section is past the chime drive pinion.

CHIME

Chime note sequences are counted on the rack (5) and quarter snail (11). Comparing this movement to other chimers, it is interesting that chime and strike are counted on a single rack. Two racks, or a rack in combination with a count wheel or locking plate, are found on most chime movements. In the Globe movement, the quarter lever (1) rests on the quarter snail and presents one of its three steps to

CHAPTER 10 - GLOBE

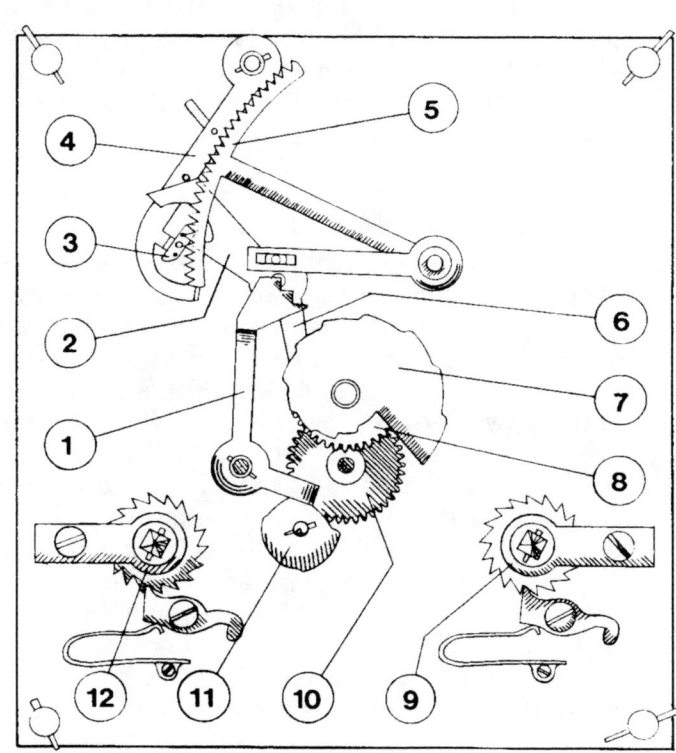

Fig. 77. Front view of Globe movement.

1. quarter lever
2. warning lever
3. gathering pallet
4. rack hook
5. rack
6. chime lifting lever
7. hour snail
8. hour wheel
9. time winding parts
10. minute wheel
11. quarter snail
12. chime-strike winding parts

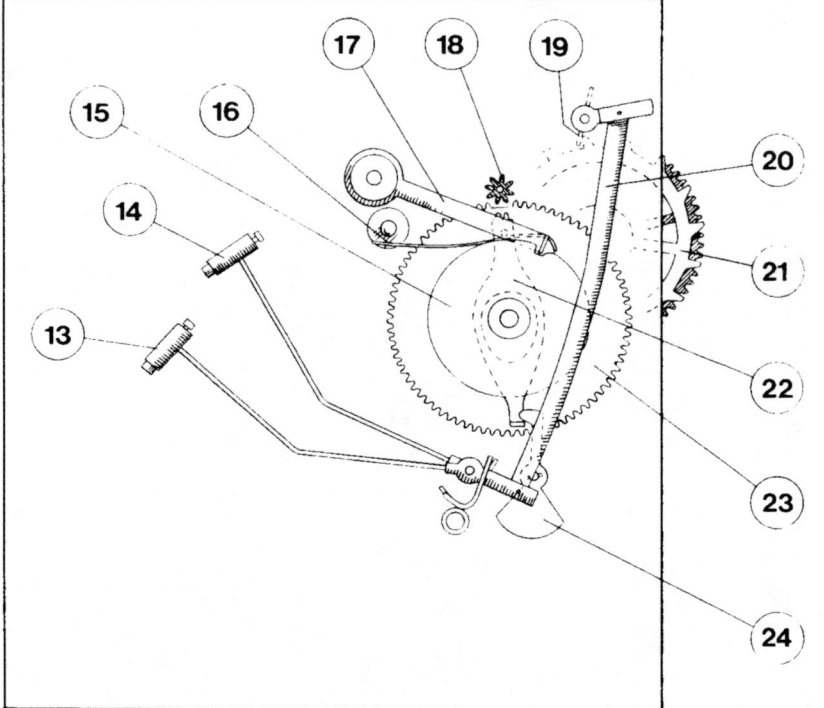

Fig. 78. Rear movement view.

13. chime hammers (4)
14. strike hammer
15. strike cam
16. sliding gear lever
17. strike silencing lever
18. chime drive pinion
19. strike hammer tail
20. strike hammer linkage
21. hammer-lifting star
22. sliding gear section
23. pin barrel wheel
24. support piece

the rack tail. The number of chime notes depends on which of the steps the rack tail has contacted. At the hour, the rack tail misses the quarter lever entirely, and falls to the hour snail (7).

Each tooth on the rack represents two chime notes. Since the gathering pallet (3) has two pins, one revolution of the pallet will account for two rack teeth or four notes in the chime melody. The chime drive pinion (18) on the rear of the movement is on the opposite end of the same arbor as the gathering pallet, so it makes one revolution for each four notes of chime melody. The chime drive pinion has eight leaves, and the pin barrel wheel has 80 teeth. This gives a ratio of 10:1. Ten revolutions of the chime drive pinion will bring the pin barrel wheel around once. In this way, the 40 notes in the Westminster chime tune are played during each hour: four at the first quarter, plus eight at the second, plus 12 at the third, plus 16 at the hour, for the total of 40.

The strike is counted as two hammer strokes per revolution of the gathering pallet, or one rack tooth per stroke. This means that at 12 o'clock the gathering pallet will first count off eight rack teeth for the chime, then 12 more for the strike, for a total of 20 teeth. The rack actually has 26 teeth.

In the Globe movement, the warning and lock pins (not illustrated) are both on the fifth wheel of the train. The warning pin is located further out from the center of the wheel. The warning run does not need to be set by the repairer, because both pins are on the same wheel.

Before each quarter hour, the chime lifting lever (6) raises the rack hook (4), releasing the lock pin. The warning pin comes to rest against the warning lever (2). A minute or two in advance of the chiming point, the rack hook has been raised enough to release the rack (5). The rack falls against one of the steps of the quarter lever or against the hour snail, depending on the time. Exactly at the quarter, the train begins to run.

CHIME CORRECTION

Most chime clocks have a self-correcting device to silence the chime until a fixed point (usually the hour) has been reached. A few movements allow the melody to play extra sequences instead, with the same result. The Globe concept is unlike either of these two. It does, however, remind us of the Sessions two-train chime featured in Chapter 7. The number of revolutions of the wheels never varies, even in a correction phase. The movement will play incorrect notes through and including the hour. Then the correct notes resume at the first quarter.

As we have seen, the Globe's chime drive pinion makes one revolution at the first quarter, two at the second, and so on. To this must be added a half revolution per strike hammer blow, at the hour. The clock may chime and strike incorrectly, and it soon reaches the end of the hour chime. From then on the clock will only strike, because the sliding gear section remains disengaged from the chime drive pinion. When the clock reaches the first quarter, the chime lifting lever raises the sliding gear section high enough to engage the drive pinion. Since the last note chimed was the last note of the hour melody, the clock now picks up the first quarter melody correctly.

11

WINTERHALDER

Two versions of a Winterhalder movement are covered in this chapter. I studied and repaired a German tubular bell movement, then was fortunate enough to receive a rod chime model of the same basic movement a few months later. The back plate of the tubular bell movement was stamped "Colonial Mfg. Co." and "No. 53". Under these words was a trademark: "Miller", the image of a clock dial, and "Germany". I thought the clock to be one of the early Colonials, with Miller as the German movement manufacturer (not Howard Miller, the American company of today).

After I published an article on the movement, Steve DeYoung of Sligh Clocks provided some background information. Although Colonial and Miller were the only names on the tubular bell movement, he said it was made by Winterhalder. The rod chime movement I received a few months later confirmed his information, because it was marked "Winterhalder" in addition to "Colonial" and "Miller". As Mr. DeYoung understood it, the parts were made in Germany, by Winterhalder, for assembly in the U.S., for Colonial. The Miller was Herman Miller, head of Colonial Mfg. Co., who teamed with Hans Winterhalder to form Miller Movement Co.

The tubular bell clock is about 6-1/2 feet tall. The single glass door reaches from above the dial down to the base of the cabinet. The movement and the tube rack are mounted on a seatboard. Five tubular bells hang from six pegs on the rack; each tube shares one peg with the next tube. The four bells for the tones of Westminster are arranged with the highest note (shortest tube) on the right, moving down to the lowest on the left. The longest tube of the five is on the far left, for the hour strike. Three weights on chains provide power for the movement. The pendulum shaft is wooden, with a light bob faced with brass. The escapement is deadbeat.

The rod chime movement is made the same as the tubular bell version except, of course, for the hammer arrangement. Chime and strike levers are the same in both movements. The rod movement needed cleaning, and a few pivots were in need of polishing. The escapement was out of adjustment, however. The removable steel pallets were worn and had been turned end for end, but not set correctly. They required adjustment to eliminate mislocking.

This chapter provides a servicing guide for the Winterhalder/Colonial movement, whether tubular bell or rod chime. We begin with an explanation of the strike and chime parts, then move on to cover basic movement assembly procedures and specifics of the tubular bell and rod chime versions.

The first thing one notices about the Winterhalder is the heavy, solid construction. In these examples, the pivots and holes show very little wear. Even stress points, such as the arbors carrying the locking cams for chime and strike, are not worn. Another feature is the design of the chime and strike levers. Much of the "action" takes place between the plates. The locking plate and locking cams are hidden from view instead of being located prominently out front, where we are used to seeing these parts. In the tubular bell version, the pin barrel and chime hammer assembly almost completely obscure the interior of the movement.

STRIKE

The strike setup is a rack and snail. Figure 79 is an overall front view of the movement which includes the layout of the strike mechanism. The rack, snail, rack hook, and gathering pallet are illustrated. Figure 80 shows the details of the strike parts. The elongated third arbor which carries the gathering pallet (6) is ground flat to receive the set screw. This arrangement means there is only one way to put on the gathering pallet. It also means that there is no reason for repairers to overtighten the set screw out of fear that the gathering pallet will slip on the arbor. The front and side views of the gathering pallet show it to be made up of two disks on an arbor. An offset pin, firmly held between the disks, gathers

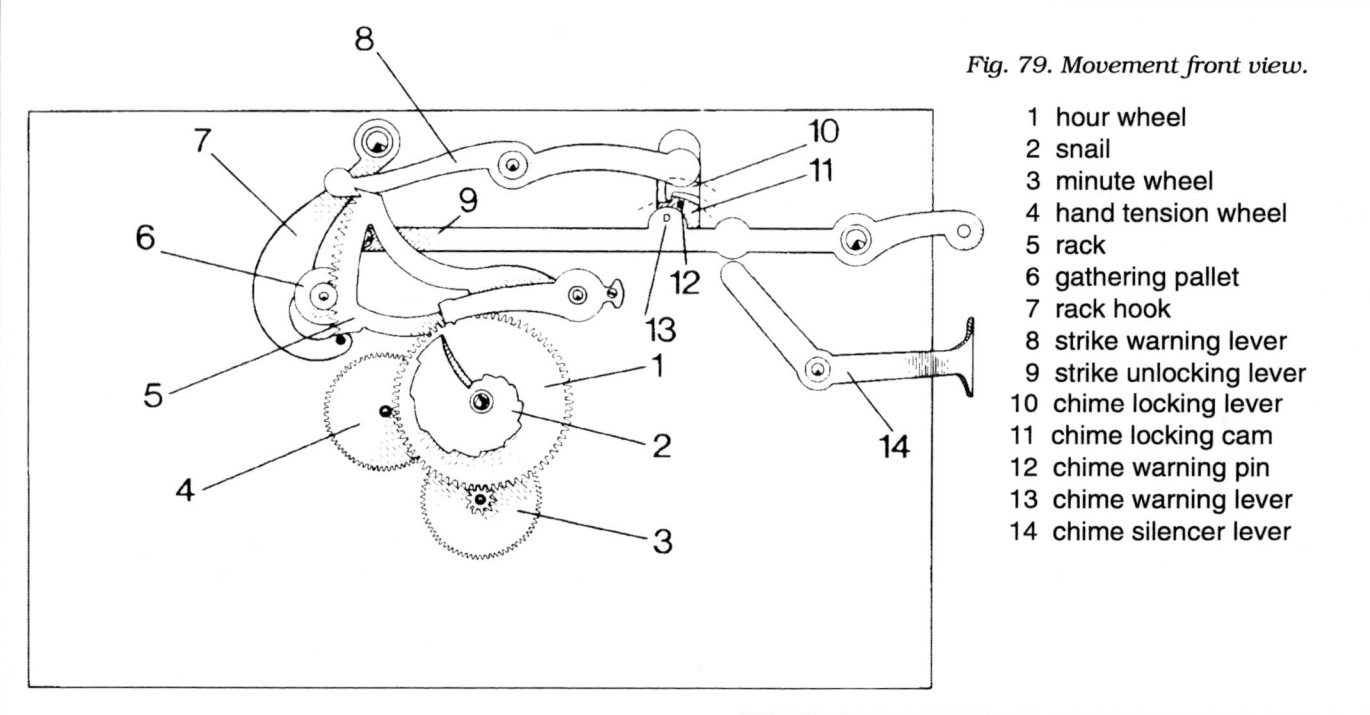

Fig. 79. Movement front view.

1 hour wheel
2 snail
3 minute wheel
4 hand tension wheel
5 rack
6 gathering pallet
7 rack hook
8 strike warning lever
9 strike unlocking lever
10 chime locking lever
11 chime locking cam
12 chime warning pin
13 chime warning lever
14 chime silencer lever

the rack. Another offset pin raises the rack hook (7) as the first pin moves the rack.

Behind the front plate, on the same arbor, is the strike locking cam (36). At the end of the strike cycle, the strike locking lever (35) drops into the notch in the cam to stop the gear train. The locking lever is attached to the rack hook and projects through a slot in the front plate, toward the cam. The strike locking cam and gathering pallet are always synchronized, because they are on the same arbor. The cam is driven on, and the gathering pallet, as mentioned above, only goes on one way. Some modern German movements also use a strike locking cam, but the arrangement is different; the cam doubles as the gathering pallet, outside the front movement plate. Also, the gathering pin is fastened to the cam. This "modern" way is much simpler and less costly to make. But the Winterhalder movement is impressive because of the strong construction. By placing the cam between the plates, the clock's designer distributed some of the strain of locking to the rear pivot. Presumably, there will be less wear on the front pivot hole as a result.

The strike hammer-lifting star and hammer arbor for the tubular movement are shown in Figure 81. As the star moves, it pushes up the hammer tail. The hammer arbor is, in turn, connected to the vertical lever which moves the strike hammer assembly on top of the movement. The drawing shows the hammer-lifting star as it raises the hammer tail during striking.

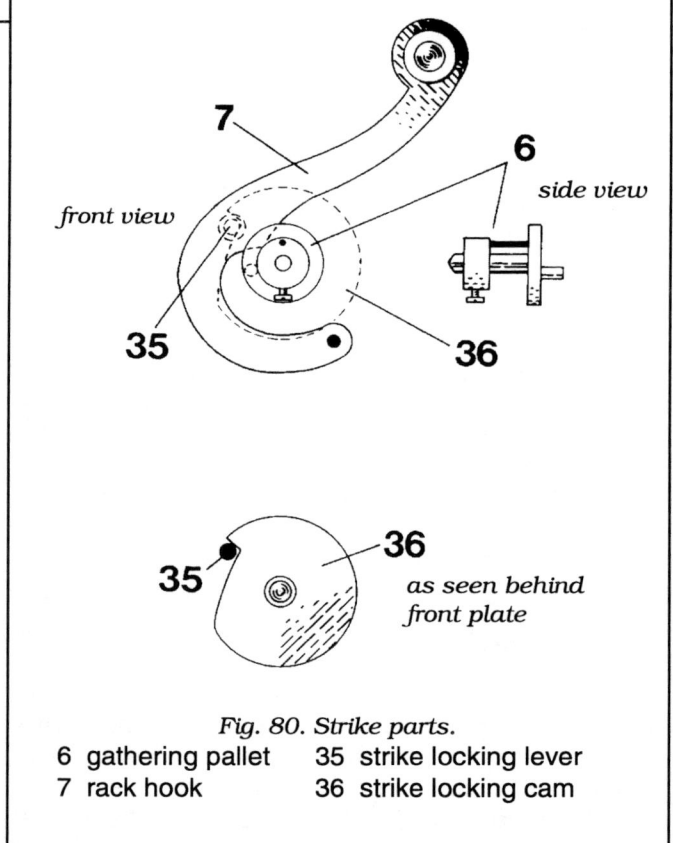

Fig. 80. Strike parts.

6 gathering pallet   35 strike locking lever
7 rack hook   36 strike locking cam

CHIME

Figure 82 shows front and side views of the chime locking wheel. The complete chime locking mechanism is drawn in Figure 83. Two cams are mounted

CHAPTER 11 - WINTERHALDER

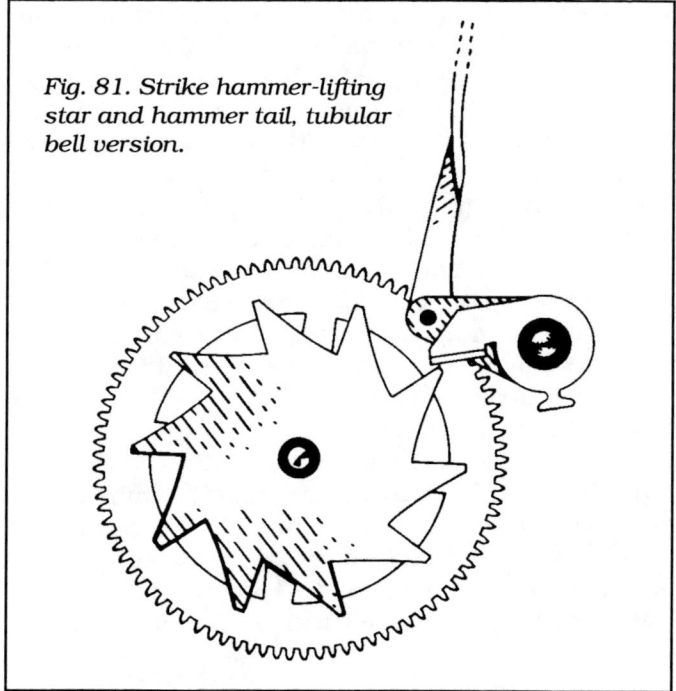

Fig. 81. Strike hammer-lifting star and hammer tail, tubular bell version.

stand the automatic chime correction feature, observe the position of the parts after the gear train completes the hour, first quarter, or second quarter. You will notice that, although the chime locking lever is engaged with its cam, the chime correction lever is raised. Figure 83 shows the parts in this arrangement. The two levers are not, after all, shaped exactly the same. The chime correction lever is made so that it cannot drop along with the chime locking lever unless the slot in the locking plate is extra wide.

The slot for the third quarter is in fact wider, allowing both levers to lock at the end of the chime sequence. The slot is shown at the bottom edge of the locking plate in Figure 83. It takes higher lift to

on the arbor, in front of the wheel and pinion. The first cam, just behind the front plate, is the chime locking cam (11). The chime locking lever (10) moves over the cam as the chime train runs. When the chime drop lever enters a slot in the locking plate (40), the locking lever catches the chime locking cam to stop the gear train. The chime warning pin (12) is mounted on the cam, and since the cam and pin move together, you do not have to set the chime warning run when assembling the movement.

The chime correction cam (37) is mounted directly behind the chime locking cam and looks identical to it. The chime correction cam works in conjunction with the chime correction lever (38). Curiously, the chime correction lever also appears to be the same shape as the chime locking lever, giving us a matched pair of levers and cams. To under-

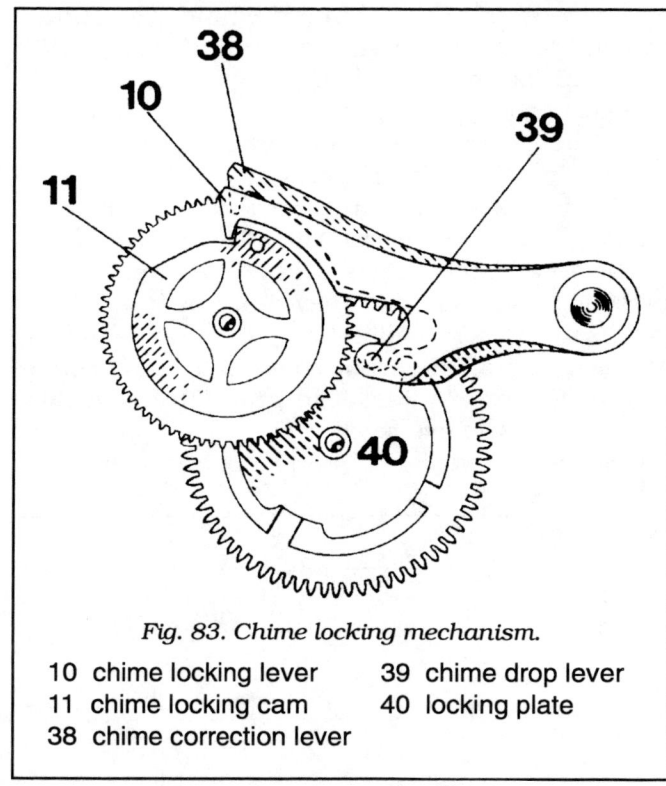

Fig. 83. Chime locking mechanism.

10 chime locking lever 39 chime drop lever
11 chime locking cam   40 locking plate
38 chime correction lever

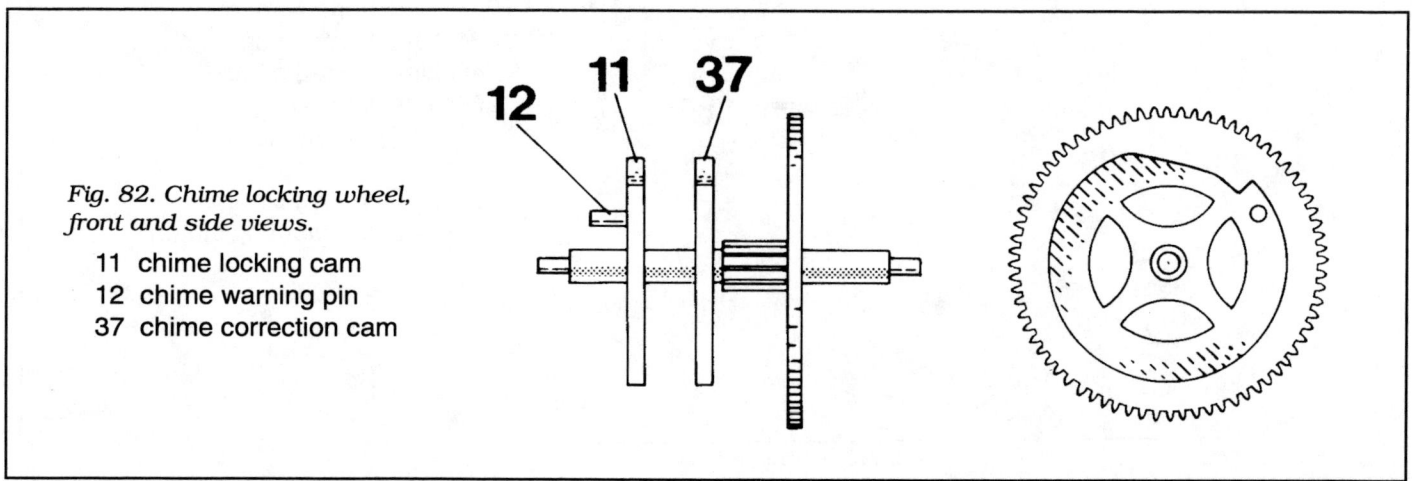

Fig. 82. Chime locking wheel, front and side views.

11 chime locking cam
12 chime warning pin
37 chime correction cam

unlock the chime correction lever. The star cam on the center arbor has one arm longer than the other three. This is for the hour.

REPAIR AND ADJUSTMENT

The basis for any overhaul is a good cleaning. Refer to Figure 79 and remove the strike warning lever (8), rack hook (7), rack (5), hour wheel (1), minute wheel (3), tension wheel (4), and gathering pallet (6). Pull the center arbor out the front, after you rotate the offset disk which holds it at the back plate. Remove the hammer assemblies, whether tubular bell or rod chime.

You may want to leave the strike unlocking lever (9) and chime warning lever (13) attached to the front plate during ultrasonic cleaning. The chime locking lever (10) and chime correction lever (38, Figure 83) are behind the front plate, attached to the same arbor. There is still plenty of room to get the chime locking wheel out and back in again. If you decide to remove the parts, it is not difficult to put them back, because there are no adjustments. The long lever (with levers 9 and 13) is driven onto the arbor, and the chime locking lever and chime correction lever slip on. They move by gravity, so there is no set-screw adjustment to worry about. At the back end of the arbor there is a brass collar with a set screw. Adjust it to provide comfortable endshake for the two levers.

Before cleaning and drying the disassembled movement parts, take a moment to inspect the chain wheels: the clicks can become sticky, and may fail unless they are individually cleaned. To accomplish this, take the chain wheels apart by removing the slotted tension spring from the arbor. In addition, the chime hammer assembly deserves careful attention during an overhaul. Do not try to clean the assembly, even ultrasonically, unless you take it apart first. You will need to polish away any dirt and rust by hand. Tightness and resistance in the hammer assembly are critical, and can stop the movement later on by increasing the load on the chime and strike trains.

To reassemble, place the arbors into the front plate. Then add the back plate, fitting each pivot carefully into place. In the rod version, make sure you install the chime drive pinion (33), shown in Figure 85. You cannot add it later unless you disassemble the movement again. Fasten the pillar nuts finger tight only. Proceed with strike adjustments first.

Strike Adjustments

Install the rack hook, gathering pallet, and rack now. Note that the gathering pallet goes on only one way. On the rear of the rod movement, you should also add the strike hammer assembly shown in Figure 84. In the tubular version, install the strike hammer tail. Turn the strike train gears by hand as you watch the action. Note that the relationship of the

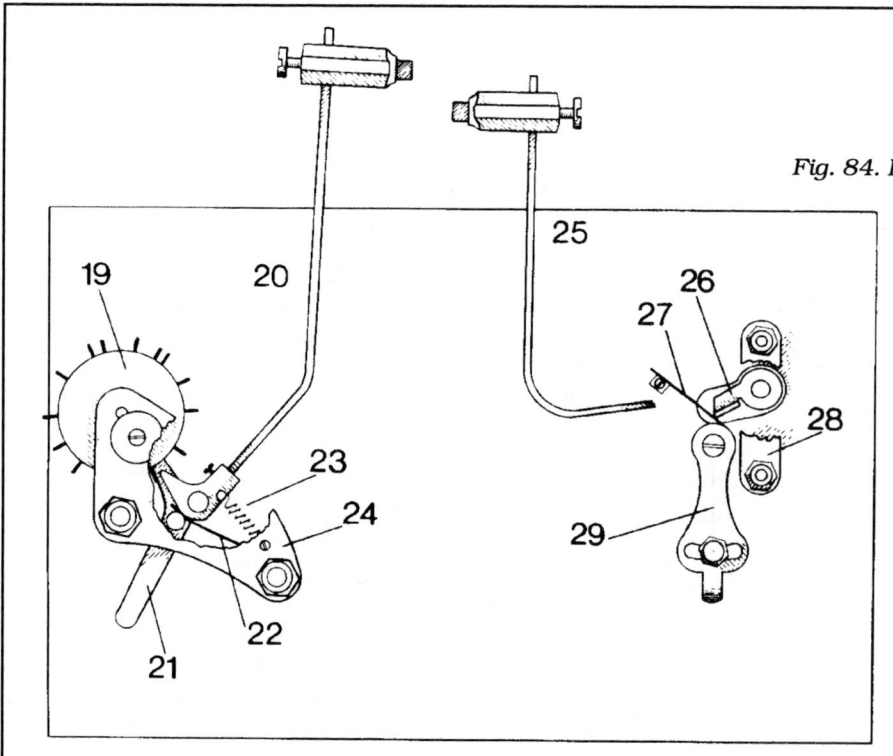

Fig. 84. Hammer assemblies, rod chime version.

19 pin barrel
20 chime hammers (4)
21 chime hammer adjustment lever
22 chime buffer springs
23 chime hammer springs
24 chime bracket plate
25 strike hammers (4)
26 strike hammer tail
27 strike buffer spring
28 strike bracket plate
29 strike hammer adjustment lever

CHAPTER 11 - WINTERHALDER

gathering pallet and the strike locking cam is fixed, because they are on the same arbor.

There are two main things to check in the strike area. First, run the gear train to the locked position and look at the strike hammer tail (26, Figure 84). If it is engaged on the hammer-lifting star as shown in Figure 81, the strike train will have to start up every time under load. Change the mesh of the second wheel and the pinion above it, until you are satisfied with the rest position of the hammer tail. Ease the plates slightly apart to accomplish the adjustment. When the strike train is locked, the hammer tail must rest between points on the star.

Second, check the strike warning pin. When the strike train is locked, the pin should be at a 6 to 8 o'clock position when viewed from the front. This will give about a half revolution of the warning wheel before the strike warning lever stops it. Separate the plates only enough to get the front pivot of the strike warning arbor out of its hole. Carefully adjust and verify the result.

Refer to Figure 84. The strike has a hammer adjustment lever (29) below the hammer assembly on the back of the movement. You can make use of it after the movement is back in the case, to change the clearance between the hammers and the chime rods. The lever applies pressure to the strike buffer spring (27), raising or lowering the four hammers together. Larger adjustments are made by loosening the set screw which holds the four-hammer unit on its arbor, or by bending individual hammer shafts.

Figure 86 shows a front movement detail view, with the hour wheel removed. The rack let-down (15) is an extension of the cannon pinion (18), providing a smooth alternative to the loud "clunk" some movements make as the rack tail drops onto the snail. In the Winterhalder movement, the rack tail comes to rest first upon the rack let-down, at the warning. Then, as the minute hand moves toward the hour release point, the rack tail slides noiselessly down the edge of the let-down.

You will not need to adjust the position of the hour wheel and snail. The hour hand is keyed to go on one way only, so it will always point to the hour to be struck. You can move the hour hand and the snail along with it, to change the hour. About the only thing you need to do is to make sure that the hour hand points exactly at a numeral as the clock chimes the hour.

Chime Adjustments- Tubular Bell
To adjust the chime train, put it completely together except for the pin barrel and hammer assembly. Referring to Figure 83, loosen the set screws on the locking plate (40) and the chime locking cam (11). Now turn the two parts with respect to each other. When the chime drop lever (39) is in a slot in the locking plate, the chime locking lever (10) should be placed to engage with the locking cam.

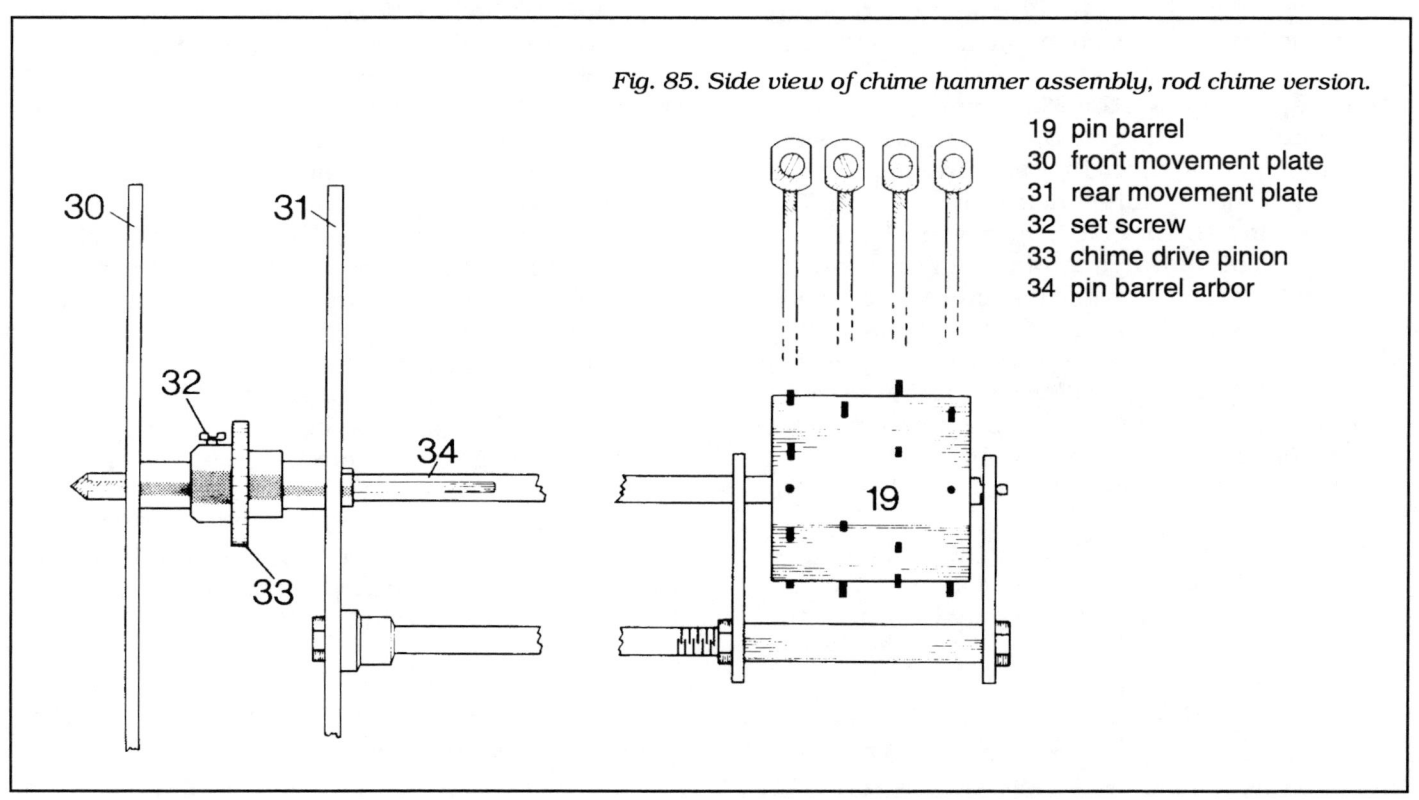

Fig. 85. Side view of chime hammer assembly, rod chime version.

19 pin barrel
30 front movement plate
31 rear movement plate
32 set screw
33 chime drive pinion
34 pin barrel arbor

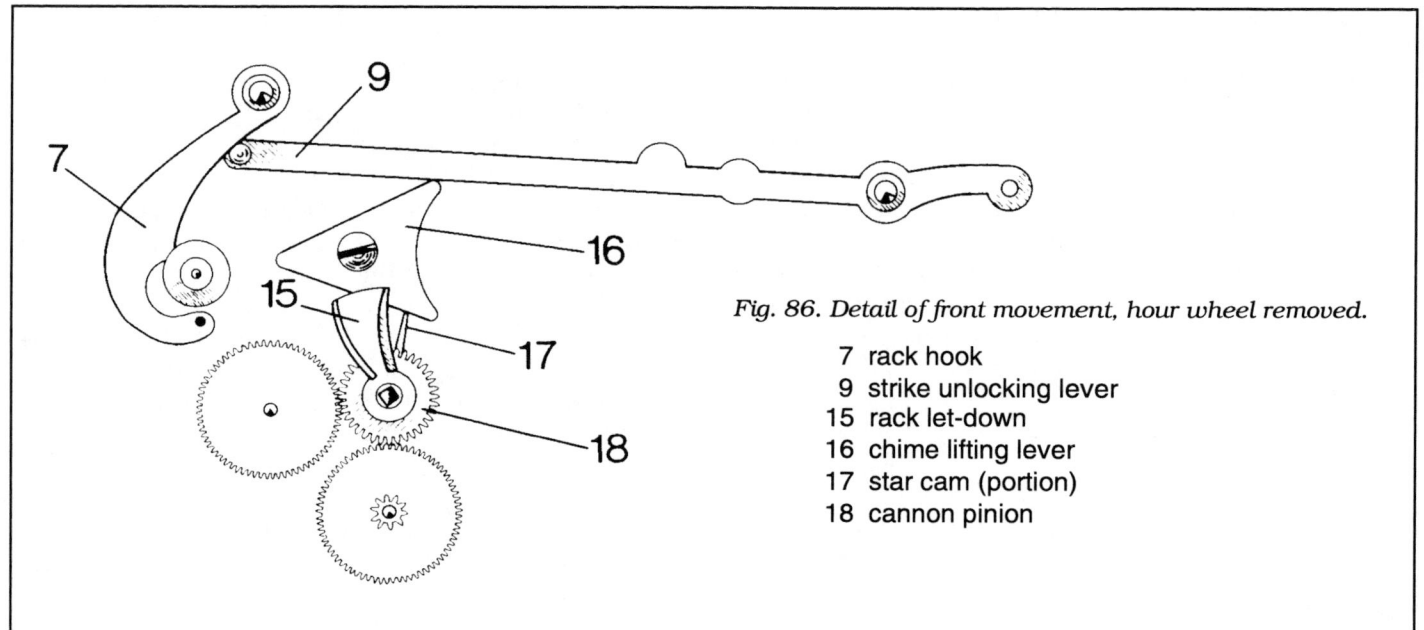

Fig. 86. Detail of front movement, hour wheel removed.

7 rack hook
9 strike unlocking lever
15 rack let-down
16 chime lifting lever
17 star cam (portion)
18 cannon pinion

Install the hammer assembly and pin barrel as one piece. Place the fully assembled movement on the seatboard and tighten the seatboard screws. Next, hook up the hammer cords. Turn the chime gears through the hour and then the first quarter. Locate the contrate wheel which drives the pin barrel, and loosen the set screw on the side. Now turn the pin barrel to operate the hammers without moving the gear train. When you have seen the four chime hammers rise and fall in succession from right to left as you face the front of the movement, stop. Tighten the set screw to complete the adjustment.

Chime Adjustments- Rod Movement
As in the tubular bell version, the chime levers are between the plates where they are difficult to see. It is fortunate that few adjustments are required. The chime warning run is fixed. Reviewing Figure 82 and 83, the chime warning pin (12) moves only a few degrees to the chime locking lever (10) at warning. There is no need to separate the plates for any chime adjustments. Refer to Figure 83 for the correct set-up. If the chime locking wheel and the wheel which carries the locking plate are synchronized, there is nothing more to worry about. You do not have to adjust the height of the chime locking or chime correction levers.

Proceed to assemble the chime hammer assembly. Figures 84 and 85 show the relationship of the parts. Insert the pin barrel arbor (34) through the hollow arbor which carries the chime drive pinion (33). Now add the rest of the hammer assembly. Note the purpose of the chime hammer adjustment lever (21). Like the arrangement on the strike side, it permits a fine adjustment to be made to all four hammers simultaneously. Using a cam, it applies varying lift to the chime buffer springs (22). This action very slightly raises or lowers the rest position of the four hammers.

Don't forget to install the chime silencer lever (14, Figure 79) on the front of the movement. It operates by lifting the chime and strike levers to the warning position and leaving them there. Chime and strike are silenced together. You don't need to adjust the chime hammer sequences until the movement is running on the test stand. The minute hand is keyed to go on only one way, and the chime train will automatically adjust itself to run the hour chime when the minute hand points upward. All that remains is to adjust the pin barrel to lift the hammers in the correct sequence.

Chime the movement through the hour, and then through the first quarter. Loosen the set screw (32) on the chime drive pinion (33), and move the pinion forward to disengage it from the third wheel. Turn the pin barrel (19) until you see the four hammers rise and fall in order from front to back. Then engage the pinion and tighten the set screw. Readjust if necessary. That's all there is to it. Chime and strike are ready for the test run.

12
JACQUES

Our subject for this chapter is a German grandfather clock marked "Jacques" on the movement and dial. The full name is Charles A. Jacques Co., of Brooklyn, N.Y. A label inside the door shows it was imported by George Borgfeldt, who was in business in New York City from the 'teens through the 1920's. The back plate is also identified with the name Bauerle, the German movement manufacturer. Several aspects of this clock make it interesting for us. It is a dual chime, but in addition to this it has a feature which permits automatic changing of the chimes. Figure 87 shows the portion of the dial where selections are made. One chime is the familiar Westminster, and the other is Trinity. If "1" is selected, the owner can change chimes when he chooses. On "2", the clock automatically alternates the two chimes every 12 hours. The clock has six tubular bells.

The tubular bells do not hang in order by length in the Jacques. To learn the correct placement of tubular bells in virtually any factory-produced clock, one should refer to Henry Fried's article "All About Chime Clocks" in the April 1982 issue of the NAWCC *Bulletin*. By checking Mr. Fried's "6A" tube hanging sequence against the Jacques clock, I found that the six hammers would produce Westminster and Trinity. His diagram shows that the tubes should hang in the following order: from left to right, with tube number 1 the shortest, the sequence is 1-3-5-6-4-2. This puts the longer tubes in the center, with the shorter ones at the ends.

CHIME OPERATION

Figures 88-90 illustrate the chime levers in the "lock", "warning", and "run" positions. At first glance, two things appear to be odd in the chime mechanism. For one, there is no chime warning lever, and consequently no warning "run" as we are used to seeing it. Secondly, the locking plate (15) looks strange indeed. It has four projections on it to correspond to the four quarters, instead of the appearance of a cam with slots as we normally see it. But the spacing of the "arms" is the same as in other locking plates: the intervals marked off are one, two, three, and four units.

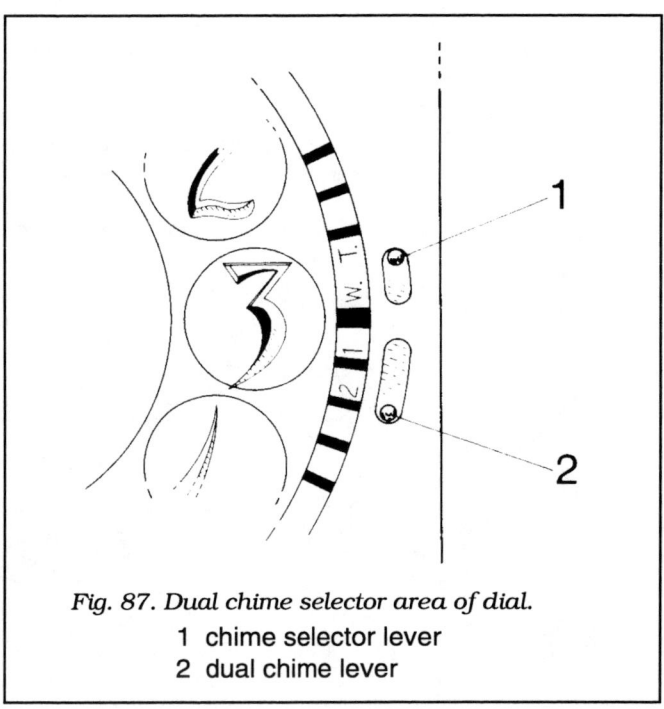

Fig. 87. Dual chime selector area of dial.
1 chime selector lever
2 dual chime lever

In the locked position in Figure 88, the chime lock pin (12) is stopped against the chime locking lever (11). Rotation of the wheels is impossible because the lock pin is on the third wheel of the train. In this position, notice that the locking plate has pushed up the chime drop lever (14). The drop lever has raised the chime unlocking lever (13). This action places the locking lever in front of the chime lock pin. Chiming cannot resume until the levers are unlocked again at the next quarter hour.

Although there is no warning run of the wheels, there is a warning action that prepares the gear train to start on the approaching quarter hour. Here's how it works. The star cam (17) raises the

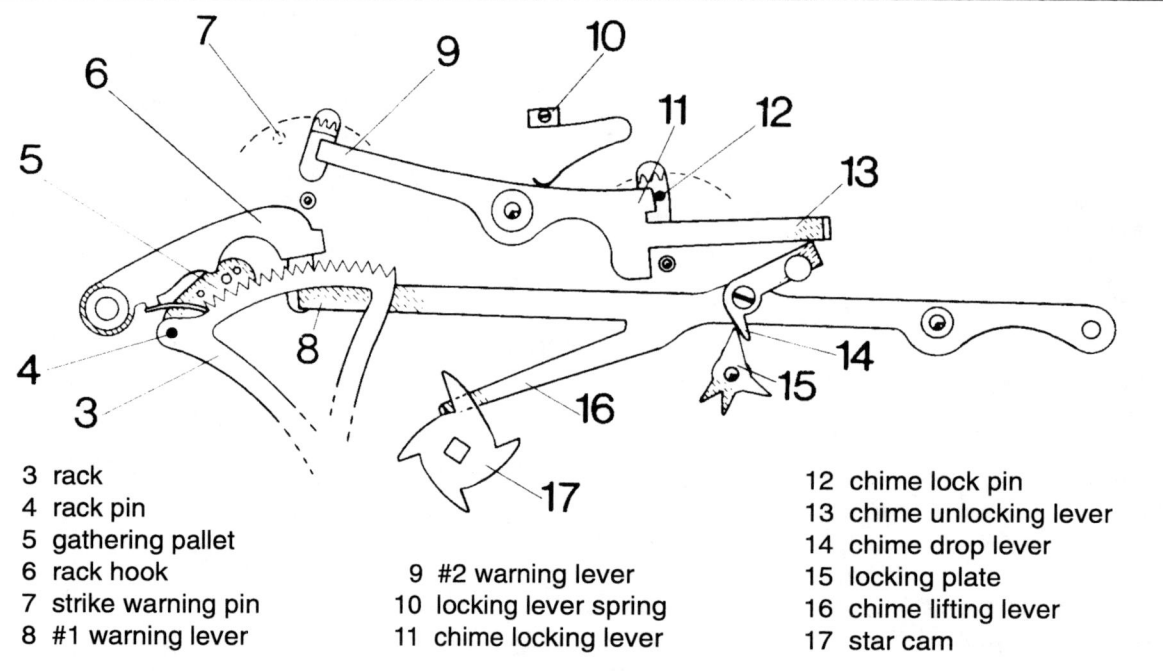

3 rack
4 rack pin
5 gathering pallet
6 rack hook
7 strike warning pin
8 #1 warning lever
9 #2 warning lever
10 locking lever spring
11 chime locking lever
12 chime lock pin
13 chime unlocking lever
14 chime drop lever
15 locking plate
16 chime lifting lever
17 star cam

Fig. 88. Chime and strike levers in the Jacques movement. The chime levers are shown in the locked position.

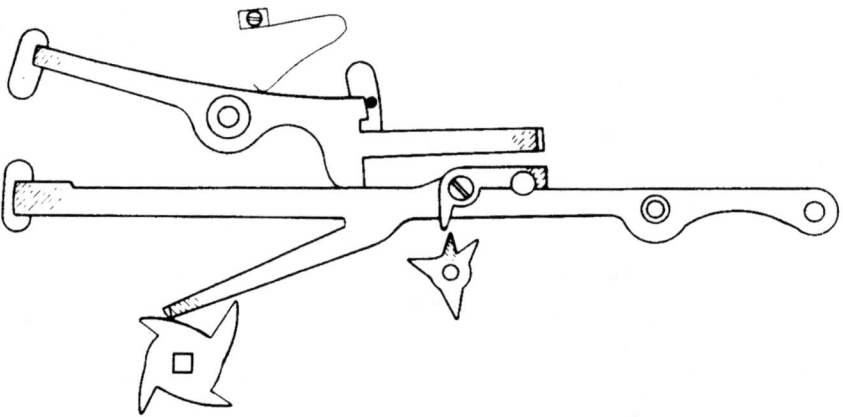

Fig. 89. Warning position of chime levers.

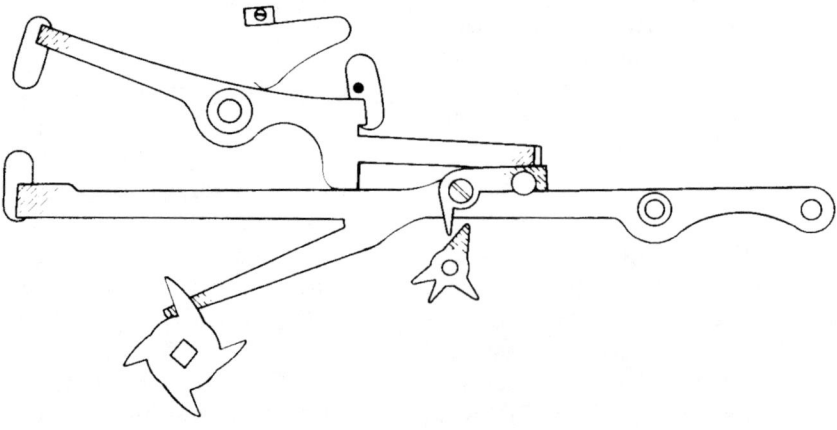

Fig. 90. Run position of chime levers.

CHAPTER 12 - JACQUES

chime lifting lever (16) and the chime drop lever (14) along with it. The drop lever goes up, then over the arm of the locking plate and comes to rest on the left side of it as pictured in Figure 89. The chime locking lever retains the lock pin during this period, and the chime train wheels have not moved.

The star cam turns far enough to release the chime lifting lever at the exact quarter hour. The chime locking lever drops away from the pin, and chiming begins. Figure 90 shows the drop lever resting in a horizontal position, out of the way. A small counterweight helps assure that it remains this way. In addition, the chime unlocking lever (13) rests upon it, aided by the spring (10). As we approach the end of the chime sequence, the arm of the locking plate pushes up the chime drop lever. In turn, the drop lever raises the chime locking lever. The slowly rotating third wheel, which holds the lock pin, runs into the locking lever on the next revolution. The chime train stops.

The chime correction feature on this movement is a simple one. The hour "arm" of the locking plate is longer than the other three arms. It requires a higher lift to bring the chime drop lever over this arm at the chime warning. The star cam also has one arm longer than the other three, and this corresponds to the hour. If the chimes have been placed out of sequence for any reason, they will automatically be corrected at the next hour. The star cam will only unlock the hour run of the chime train at the hour, with the minute hand pointing straight up.

Chime silencing is equally straightforward. The chime silencer lever (not shown) pushes up the chime unlocking lever and holds it. The chime locking pin and lever remain together, so there is no chiming. The lever pierces the dial at "6".

The design of the Jacques chime train brings a few problems with it. The chime lock pin tends to stick to the chime locking lever unless the surface is polished and angled properly. Because the locking pin is on the third wheel, it has a lot of power behind it. To smooth out the action, the lever should be angled in such a way that it never pushes the lock pin and third wheel against their counterclockwise direction of rotation. The "pushing" increases the risk that the lever will not fall away from the pin to start chiming. Even with the spring (10) helping to assure that the chime locking lever drops away to permit chiming to begin, there are still problems with sticking. This means that the angle and surface finish of the chime locking lever are all-important. It is unfortunate that the chime lock pin is on the third wheel of the train, which starts up so slowly. Any hesitation will result in a chime train that stalls out frequently.

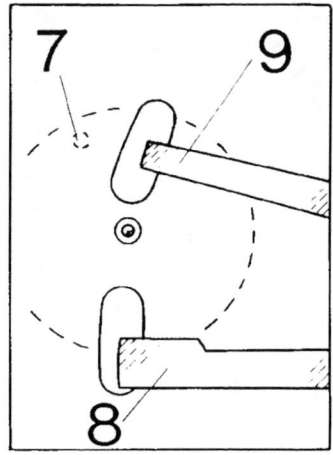
a lock position

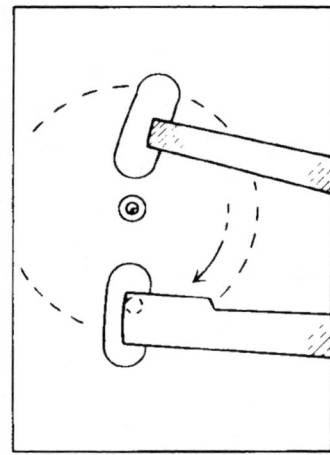
b warning position

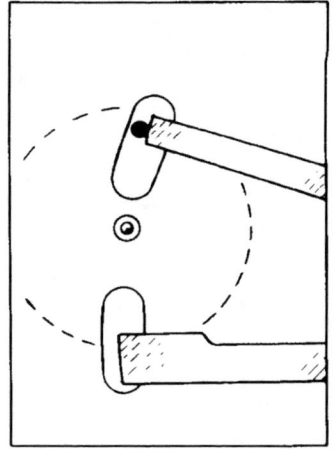
c position of strike levers during chime run

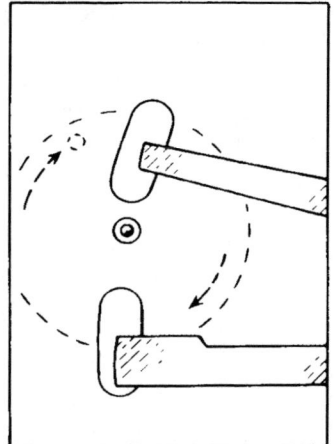
d run position

Fig. 91. Strike levers.
7 strike warning pin
8 #1 warning lever
9 #2 warning lever

STRIKE OPERATION

The strike levers are drawn in Figures 88 and 91. We are looking at a rack and snail design. The gathering pallet (5) locks on the rack pin (4) to hold the strike train at rest. This is similar to the Herschede tubular bell movement design for the strike train. But there is something a little different about the Jacques. There are two strike warning levers, (8) and (9). This seems confusing at first.

Figure 91a shows the locked, or rest position of the levers and the warning pin. The rest position of the warning pin is not terribly critical. Any position between 11 o'clock and 3 o'clock is satisfactory. Don't worry about conflict with the #2 warning lever (9). It is out of the way at this point. Moving to

Figure 91b, check the warning position for the strike. The #1 warning lever (8) always comes into play at this time, as the hour chime warning occurs. The high lift of the star cam raises the chime drop lever up and over the long arm of the chime locking plate, as we saw earlier. The lifting action also raises the #1 warning lever high enough to lift the rack hook (6, Figure 88). The rack (3) drops, and the gathering pallet slides off the rack pin. The warning pin (7, Figure 88), mounted on the third wheel, rotates clockwise until it hits the raised #1 warning lever.

Exactly at the hour, chiming begins. The #1 warning lever releases the warning pin. The wheel and pin rotate around to be stopped by the #2 warning lever, which has pivoted upward to block striking as shown in Figure 91c. In fact, this is the only function of the #2 warning lever. It prevents striking whenever the clock is chiming.

At the end of the hour chime run, the #2 warning lever pivots downward, releasing the warning pin. Striking now begins. Figure 91d illustrates the "run" position for the strike. The #1 and #2 warning levers are both clear of the warning pin.

If the chimes are silenced, the #2 warning lever is pushed permanently down and out of action. The #1 warning lever functions as before, but striking begins immediately at the hour. There is no chiming to wait through, so the strike begins at once. Only the #1 warning lever is involved. Silencing the strike simply involves moving the hour strike silencer lever, located at the numeral "9" on the dial, to "silent". The lever prevents the rack from moving, thereby keeping the strike train at rest.

DUAL CHIME SELECTOR

Figure 87 shows the chime selector area of the dial. The chime selector lever allows the choice of "W" or "T", Westminster or Trinity. However, this selection can only be made when the dual chime lever is on "1", which stands for single chime. If the lever is on "2", the clock is set to automatically switch from one chime to the other twice per day. The owner cannot determine which chime is played at any particular time. To regain control, he moves the dual chime lever back to "1". Then he can select either Westminster or Trinity to be played until he wants to change it again.

We must go behind the dial to examine the work-

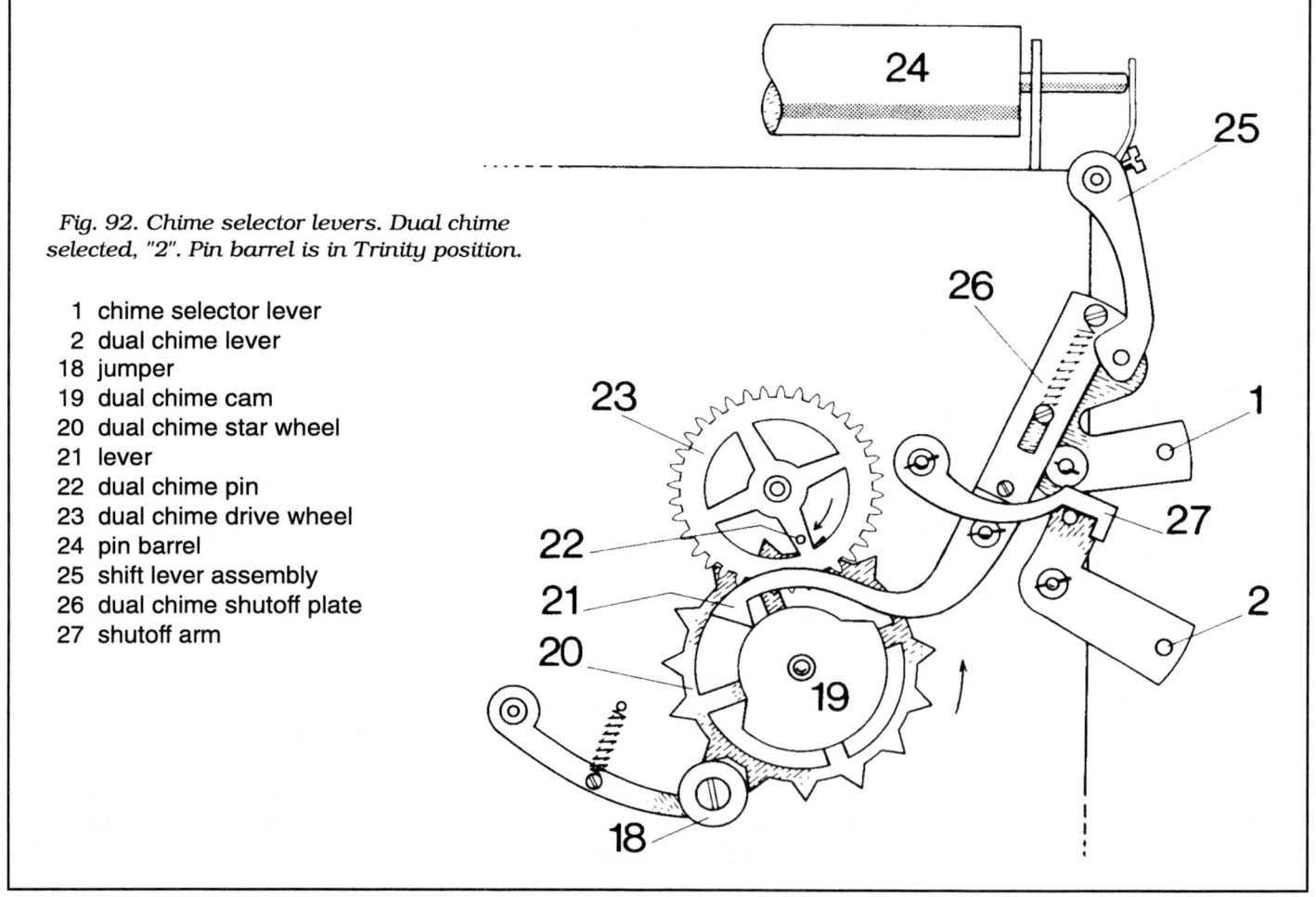

Fig. 92. Chime selector levers. Dual chime selected, "2". Pin barrel is in Trinity position.

1 chime selector lever
2 dual chime lever
18 jumper
19 dual chime cam
20 dual chime star wheel
21 lever
22 dual chime pin
23 dual chime drive wheel
24 pin barrel
25 shift lever assembly
26 dual chime shutoff plate
27 shutoff arm

CHAPTER 12 - JACQUES

ings of the chime selector mechanism. Refer to Figure 92 for an overall view of the dual chime selector, located on the right front portion of the movement. Figures 93 and 94 show the same chime selector levers in different positions. Figure 95 illustrates the mechanism on the back of the dial which prevents the user from selecting chimes when the clock is on automatic "2".

Dual Chime Position

When the dual chime selector is set on "2", the mechanism appears as in Figure 92. The dual chime drive wheel (23) is driven off the motion work. It has a pin (22) which contacts the dual chime star wheel (20), moving it one tooth space at a time. The jumper (18) holds the star wheel in each successive position. The purpose of all this is to move the dual chime cam (19). The cam has two surfaces, each covering half its circumference. The smaller diameter segment corresponds to Trinity; the other is for Westminster. We'll see why in a moment.

The cam acts on the lever (21). In turn, the lever either pushes the pin barrel to the left, against spring pressure, for Westminster, or allows the pin barrel to return to the right, for Trinity. In the illustration, the lever is in the Trinity, or rest position. Remember that the cam turns gradually, step by step. The illustrations show the cam in the middle of each position, for clarity.

Figure 93 shows the lever in the middle of the Westminster phase. The entire lever has pivoted clockwise. As a result, the dual chime shutoff plate (26) applies pressure to the shift lever assembly (25). The pin barrel (24) is pushed to the left, bringing the Westminster pins to bear on the hammers. As the cam turns around to the Trinity position, the lever goes back as shown in Figure 92, releasing the pressure on the shift lever assembly. The pin barrel moves back to the right, under spring pressure. Now the Trinity set of pins will lift the hammers.

Single Chime Position

When the owner moves the dual chime lever in the dial to the "1" position, he regains control of chime selection. He can change chimes whenever he wishes, but they no longer change automatically. Figure 94 shows the Trinity chime selected manually. What concerns us here is the means for cancelling out the automatic changing of chimes.

As the dual chime lever is moved up to "1", it pushes on the shutoff arm (27). The arrows in Figure 94 show what happens. The shutoff arm shifts the dual chime shutoff plate (26) upward, against spring pressure. The top end of the shutoff plate is now too high for it to touch the shift lever assembly

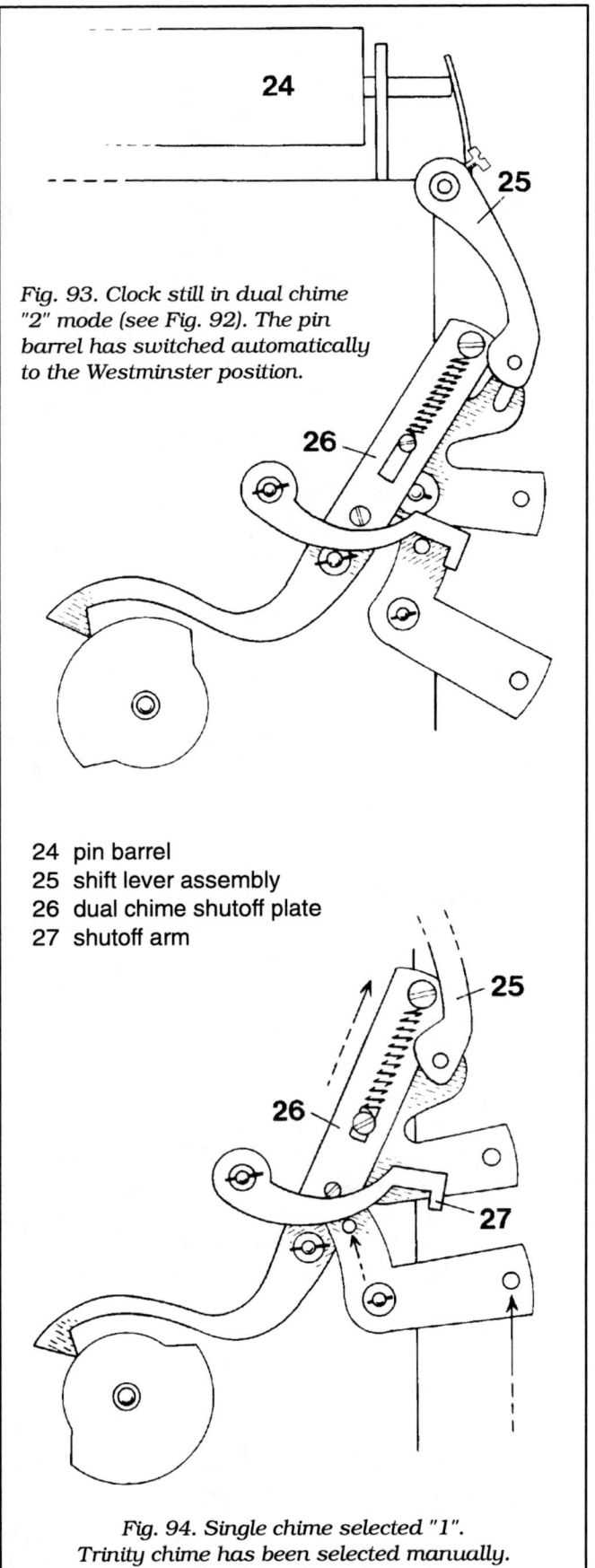

Fig. 93. Clock still in dual chime "2" mode (see Fig. 92). The pin barrel has switched automatically to the Westminster position.

24 pin barrel
25 shift lever assembly
26 dual chime shutoff plate
27 shutoff arm

Fig. 94. Single chime selected "1". Trinity chime has been selected manually.

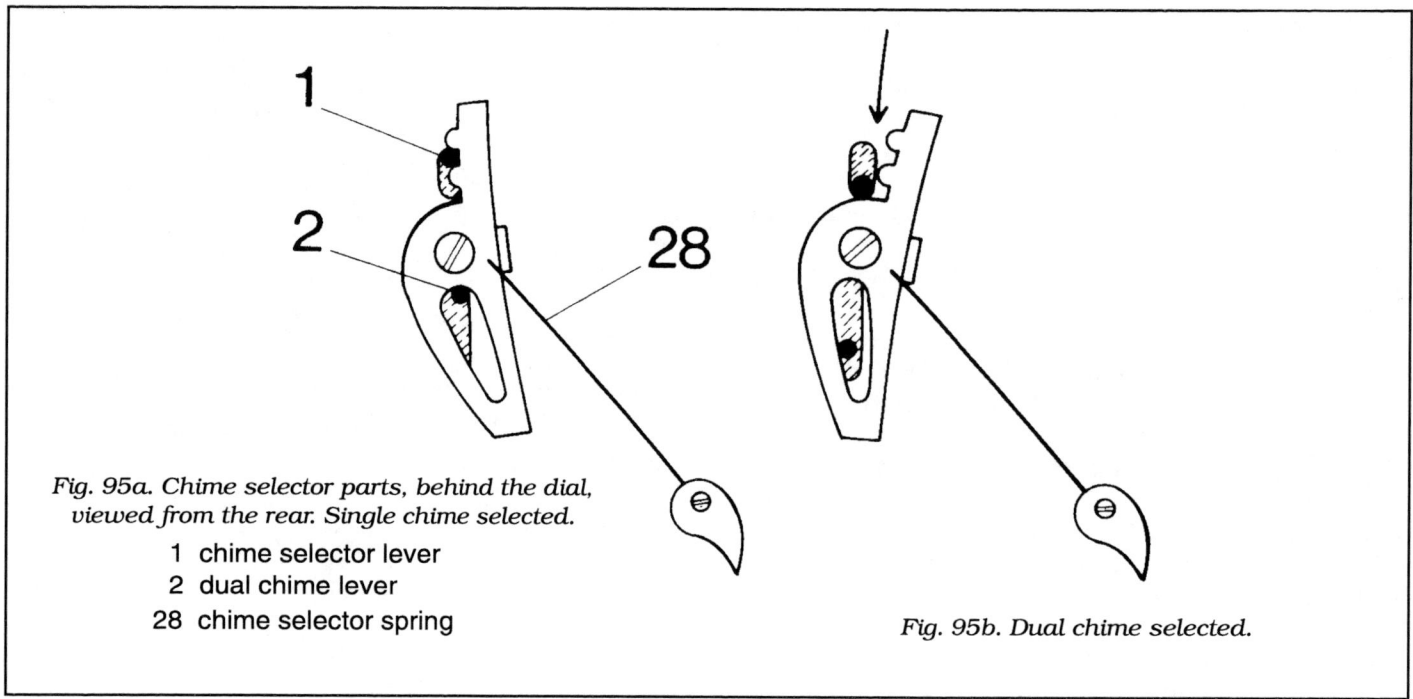

Fig. 95a. Chime selector parts, behind the dial, viewed from the rear. Single chime selected.
1 chime selector lever
2 dual chime lever
28 chime selector spring

Fig. 95b. Dual chime selected.

(25). This means that although the dual chime cam keeps moving the lever back and forth each 12 hours, the lever cannot have any effect on the pin barrel. Instead, the chime selector lever determines which chime is played. It is directly connected to the shift lever assembly.

DIAL PARTS (FIG. 95)

Now it is time to explain how the chime selector lever is able to change chimes when single chimes "1" have been selected, yet unable to do so when dual chimes "2" are automatically changing chimes each 12 hours. The two views in Figure 95 give the answer.

The parts shown in the figure are mounted to the back of the dial. The chime selector lever (1) and the dual chime lever (2) are viewed from the back—a "movement's eye view". The purpose of these chime selector parts behind the dial is to render the chime selector lever inoperative when the clock is on dual chime. In other words, the clock does not permit manual selection when on "'automatic" (dual).

Figure 95a shows the single chime position. The dual chime lever is "up" for single chime. A spring (28) pivots the unit counterclockwise, keeping pressure against the chime selector lever (1). The owner can move the selector from Trinity to Westminster at will, with the spring assuring that the selection is maintained.

Moving to Figure 95b, the dual chime lever has been pushed to the "down" position. It is here that the mechanism must cancel out the effect of the chime selector lever (1). Remember we don't want the lever to work; the clock is now on "automatic" and out of the owner's control. When the dual chime lever is pushed down, it moves in a slot to pivot the unit clockwise, against spring pressure. The arrow in the figure shows that the chime selector lever is no longer held in either position. If the automatic selection is Trinity, the lever will not stay down if it is pushed. If the selection is Westminster, you cannot push the lever up to stay.

13
JUNGHANS

This chapter describes two Junghans chime models. We begin with the movement illustrated in Figures 96-98. Another model, the B-10, is covered on page 84 and shown in Figure 99.

A JUNGHANS CHIME AND STRIKE

The Junghans chime movement shown in Figures 96-98 carries no model number. It has a removable lower front movement plate to permit access to the three mainspring barrels. The three mainsprings must be *fully* let down before the lower front plate or the winding mechanisms can be removed or loosened.

The chime train is locking plate controlled. The movement is fitted with an automatic chime correction device which prevents the hour chime from operating except at the actual hour. The strike is a rack and snail type.

Figures 96 and 97 show the left and right sides of the front plate, respectively. The parts on the left are strike, and those on the right are chime. The only part shown on both drawings is the locking plate. If the two drawings are superimposed on the center arbor, they form a picture of the entire front plate. Figure 98 is a detail of the chime levers at the end of the third quarter chime, with the chime correction mechanism in effect.

Mainsprings and Winding Mechanisms

The chime and time mainsprings are wound by auxiliary mechanisms. The winding arbor has a small gear on the end, to turn a click wheel which has a greater number of teeth. These gear down the ratio for winding: more turns with less strain. In addition, the mechanisms allow the time and chime keyholes to be evenly spaced on the dial. The strike is directly wound without an auxiliary mechanism.

Repairers often neglect to clean and inspect the mainsprings and winding mechanisms. Poor performance is the inevitable result. Sometimes there is damage to barrel teeth, wheels, and pinions from "sudden" failures which could easily have been avoided. Always check for cracked mainsprings, damaged click wheels, and weakened click springs. These visible warning signs usually come before serious damage occurs. If a clock is received for repair because it chimes only five days on a winding, it's just as important to look after the springs as it is to clean and polish the pivots. It's amazing that some repairers wonder why a clock doesn't perform well with bad springs.

Pin Barrel

The pin barrel is mounted on the rear of the movement. It is easier to adjust than most, because the manufacturer has marked the quarter hour chime point for you. Here's how it can be adjusted easily. The pin barrel consists of two large brass disks. A row of steel pins is driven into each flat surface, making four in all. Each row operates one of the four chime hammers. To adjust the pin barrel, first chime the movement through the hour and then the first quarter. Loosen the set screw on the pin barrel arbor, and turn the pin barrel by itself. It moves clockwise when viewed from the back. Watch the arrow marked "1/4" as you turn clockwise. Just as the arrow passes by the hammer levers, stop. This is the end of the first quarter chime: tighten the set screw. Four hammers have just been raised in order to play the four descending notes of the first quarter chime.

Chime Train

Locate the long curved lever above the right side of the front movement plate. This is the chime correction lever (25). It is not a chime locking lever intended to stop the gear train after each chime. It operates only after the third quarter chime. The lever has a brass wire spring (26) wound along its arbor. The spring is secured to a screw head in the movement pillar. The spring applies pressure to move the tip of the lever up, not down. The spring

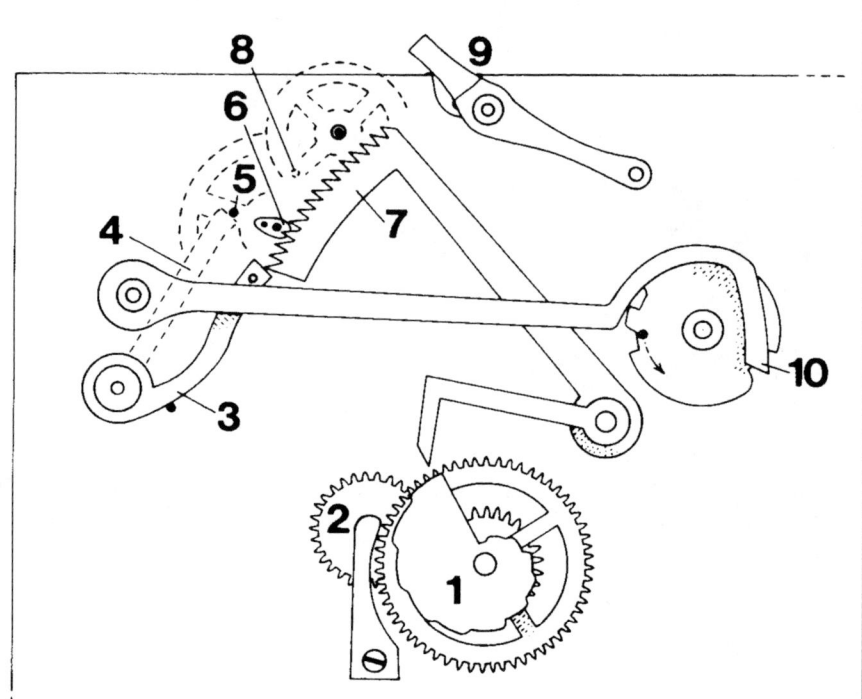

Fig. 96. Junghans chime movement, principal strike parts on left front side.

1. hour wheel and snail
2. minute wheel
3. rack hook
4. strike locking lever
5. strike lock pin
6. gathering pallet
7. rack
8. strike warning pin
9. strike warning lever
10. strike unlocking lever

must be in good condition to apply firm pressure.

The chime train stops on the chime locking cam (19). To operate the chime train, the lifting piece (14) moves the chime unlocking lever (17) upward. In turn, it pushes up the chime locking lever (21) and chime lock pin (20). After the unlocking occurs, the chime warning pin (24) moves counterclockwise to the chime warning lever (16). Figure 97 shows the warning pin at a 1 o'clock position for clarity in the drawing. The actual rest position of the warning pin should be at an 8 o'clock orientation. The warning lever is at about 6 o'clock, which makes the warning run relatively short. It is necessary to set the warning this way because of the chime correction mechanism.

Chime Correction Mechanism

Most of the adjustment problems are concerned with the chime correction device. The mechanism itself operates as follows. The locking plate (15) has a deep slot for the third quarter locking point. Figure 98 shows the chime drop lever (18) in this slot. The action pulls the chime locking lever (21) lower than for the other quarters. The chime correction guide arm (22) is pivoted on the lever, and it pulls down on the chime correction lever (25). The lever moves down just enough to catch the chime warning pin (24) as it comes around.

Figure 98 shows the chime correction mechanism locked in position. The angled notch in the chime correction lever forces it even lower as the warning pin enters the slot. Unless the lever is spring-loaded up (not down) none of this chime correction will occur.

The cannon pinion (12) has four lift pins (11). Only the hour pin has enough lifting action to unlock the chime train for the hour. Figure 97 shows the back (riveted) end of each pin. The acting part of the hour pin is a built-up steel pin of half round shape, larger than the other three pins.

Note in Figure 98 that the chime train is not stopped on the chime locking cam. The locking pin is in the slot but not touching the edge. The 8 o'clock warning pin position noted above is critical. If the pin is out of position, one of two problems will occur. The chime correction lever may miss the pin, allowing the chime train to lock as usual on the cam. There will not be any chime correction in this case. Or the chime correction lever may catch the pin too soon, locking the train early. On start-up, the outer rim of the locking cam will not be able to get under the locking pin quickly enough. The train will lock again without chiming the hour.

Adjusting The Chime Levers

The drop and locking levers (18 and 21) form a unit riveted together. By opening or closing the angle, you can change the amount of locking action. If the angle is closed too much, you lose the chime correction feature. The chime locking pin bottoms out in the locking cam, making it impossible for the drop lever to go all the way into the deep slot in the locking plate. If the angle of the levers is opened too much, you lose the locking action of the pin against

CHAPTER 13 - JUNGHANS

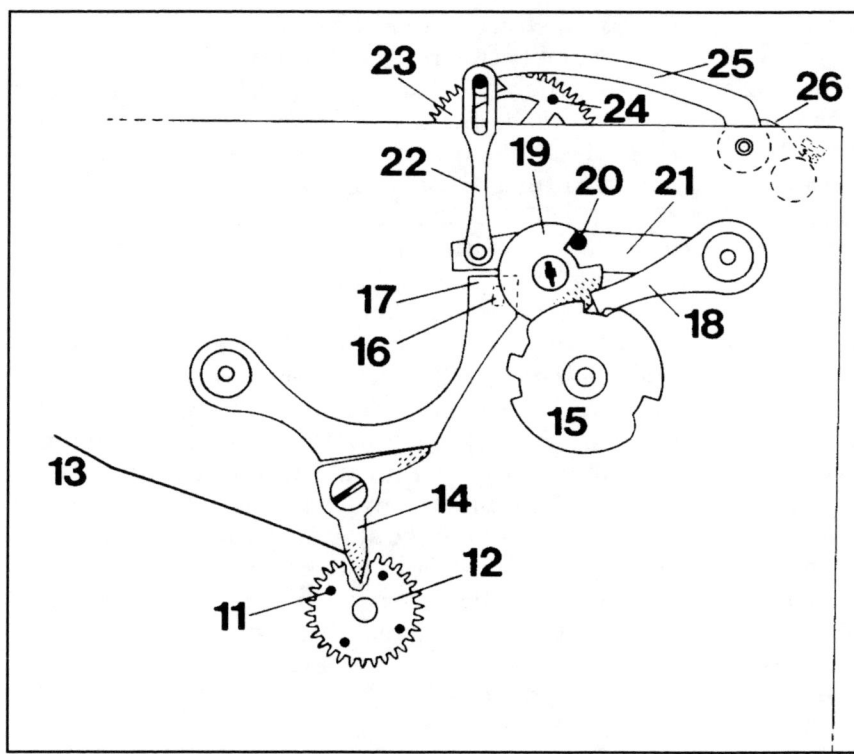

Fig. 97. Chime parts from right front portion of movement.

11 chime lift pins (4)
12 cannon pinion
13 lifting piece return spring
14 lifting piece
15 locking plate
16 chime warning lever
17 chime unlocking lever
18 chime drop lever
19 chime locking cam
20 chime lock pin
21 chime locking lever
22 chime correction guide arm
23 chime warning wheel
24 chime warning pin
25 chime correction lever
26 spring for correction lever

the slot in the locking cam.

It is usually a mistake to start bending the chime correction lever (25) in an attempt to make the chime correction feature work. One problem is that there is very little operating clearance between the tip of the lever and the chime warning pin as it moves past. If the lever itself is bent lower, the chime warn-

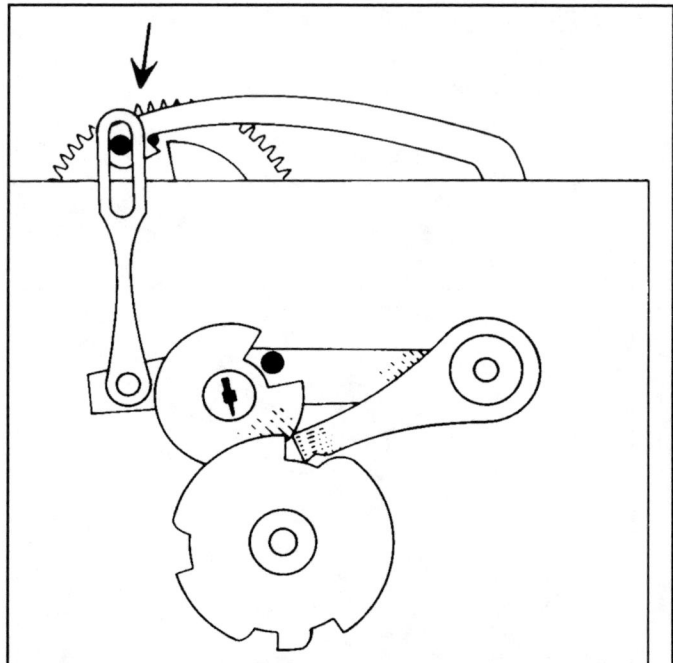

Fig. 98. Chime mechanism in "correction" phase following the third quarter chime. The chime correction lever catches the warning pin (at arrow). This stops the gear train. The locking cam and pin are not engaged.

ing pin catches it at each revolution—even during the chiming. If the lever is too high, it cannot catch the warning pin when it is supposed to do so. Instead of bending the chime correction lever, a small adjustment can be made by bending the brass pin driven into the tip. Bending the pin down sets the lever higher; bending it up sets the lever lower.

I suspect that a lot of time is wasted in adjusting the chime correction lever. It often happens that the drop/locking lever adjustment described above is really the problem. It is a good idea to do that adjustment first, so the chime correction feature can work. Often, the chime correction lever has been broken from repeated bending, then soldered together. The drop and locking levers have also been moved a great deal, with more solder applied.

When the chime correction feature operates, it is important for the warning pin to remain locked in the angled notch of the lever. Verify that none of the first three quarters will have enough lift to unlock the device. It will be close each time, which is one of my complaints with the movement. Because of the angle of the notch in the chime correction lever, the hour lift actually backs up the chime train to accomplish the unlocking.

Strike Train

Fortunately, the strike train is not unusual or problem-ridden. Figure 96 shows the positions of the gathering pallet, strike locking lever and pin, and strike warning pin. The strike unlocking lever (10) is operated during the hour chime by a pin on the locking plate. The pin is shown in the drawing as it moves counterclockwise toward the lever.

One point worth noting concerns the strike warning lever (9). It is spring-loaded clockwise, which keeps the left side up. During the hour chime, the chime locking lever pushes the strike warning lever downward against spring pressure, to be in position when the strike train is unlocked for the warning. If the spring is set up backwards, the strike warning pin will not be able to get past the warning lever at any time.

JUNGHANS B-10 CHIME

The B-10 has a different arrangement of levers than the Junghans model described earlier in this chapter. The B-10 seems to be a cheaper model, because the front plate is all one piece. On better movements of this type, the winding arbors are carried in a removable lower front movement plate. The plate reduces repair time during reassembly and servicing, because the barrels can be removed without disassembling the entire movement.

The front view of the movement in Figure 99 shows the major parts that are located on the outside of the front plate. The drawing also shows the connection between the chime and strike trains and the unusual way the chime mechanism starts up the strike each hour. The slowly rotating strike cam (14) raises the strike unlocking lever (5) each hour, which in turn unlocks the rack hook (13) and the rack (12). The strike warning lift arm (11) attached to an arbor in the chime train, moves the strike warning lever (4) into position.

The inset view in Figure 99 shows the between-the-plates parts which control the chime counting. The locking plate (9) has four slots to mark the quarter hour locking points. The chime count lever (8) rides on the locking plate to count the sequences. The chime drop lever (6) and chime cam (7) serve to prevent the chime train from locking prematurely as the chime sequence starts up each quarter hour. There is nothing unusual in these parts, except perhaps that there is no chime correction mechanism in this movement. If the chime sequence is upset for any reason, the clock does not automatically bring the chimes into synchronization with the minute hand position.

1 chime lifting lever
2 chime warning lever
3 chime locking lever
4 strike warning lever
5 strike unlocking lever
6 drop lever
7 chime cam
8 chime count lever
9 locking plate
10 lifting piece return spring
11 strike warning lift arm
12 rack
13 rack hook
14 strike cam

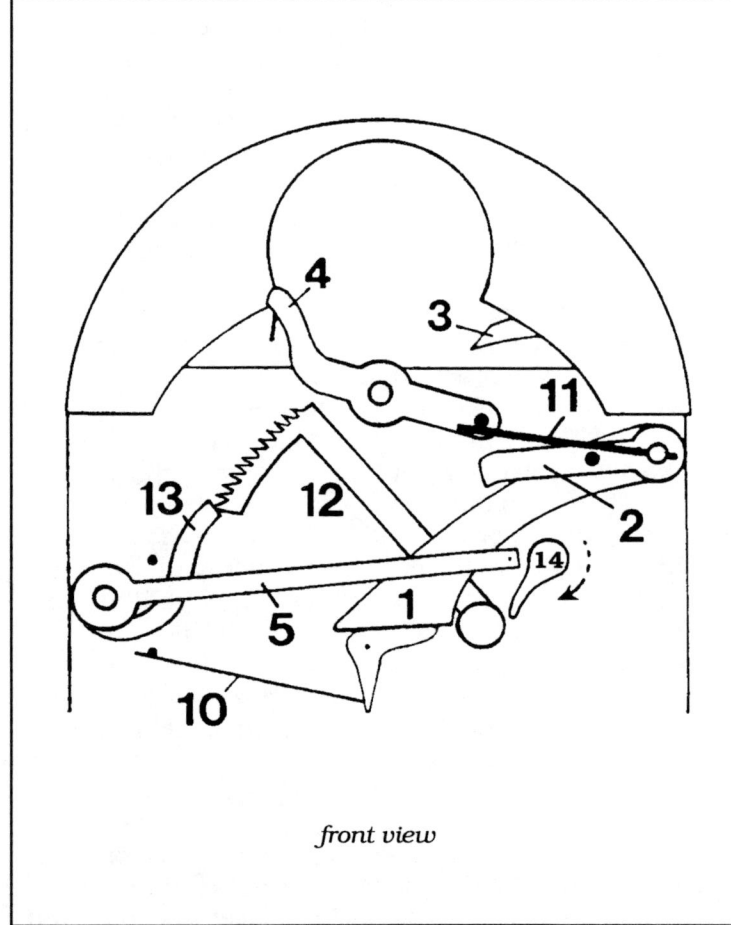

front view

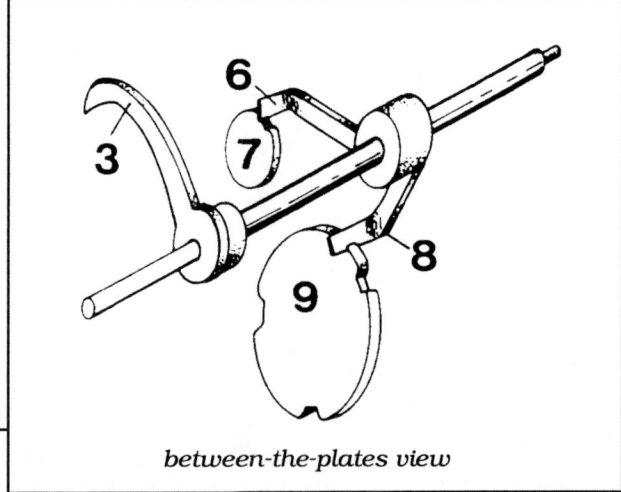

between-the-plates view

Fig. 99. Junghans B-10 chime movement levers.

14
HERMLE

German manufacturers produce almost all of today's chime movements. Most of the U.S. clock companies use them, and as thousands are sold each year you can look for steadily increasing numbers appearing for repair. The Hermle chime movements are the most numerous. There are several versions of the design: spring and weight driven, pendulum, floating balance, platform balance, and even one with a quartz time train in a mechanical chime movement. They go into a variety of wall, mantel, and grandfather clock cases. In this chapter we consider the basic Hermle chime mechanism found in the 340-020 and 1050-020 mantel chime movements and the 451 and 1151 floor clock movements.

Repairers differ in their opinions of what kind of servicing the Hermle movements require. Some of the movements have needed major work after only a few years. Clockmakers who gained most of their experience on durable antique movements find this difficult to accept. Customers can't accept it, either. The cost of restoration, when compared to the relatively low purchase price of the clock, may result in a rejected repair estimate. The repairer is pushed in one of two ways: the repair of the worn movement, or replacement with a new one. Repair is favored by many hobbyists and some professional repairers who want to do the work. As an alternative, replacement is a fast, cost effective repair that is usually beneficial to the repairer, who reduces his or her work time, and to the customer, who receives a new mechanism which will probably perform well for years.

Any Hermle movement that is to be repaired must be given proper attention. It is possible to save repair time by removing the mainspring barrels (if spring driven) and cleaning the movement ultrasonically. But most of the movements will require disassembly anyway, for pivot and bushing work. Mainsprings should be removed from the barrels, inspected, and cleaned. The notion that most of the movements can be cleaned while assembled, then quickly oiled and re-cased, is a fantasy.

IDENTIFICATION AND DATING

Many Hermle movements were stamped "Hermle" or "FHS" along with the model number, such as 340-020. Most Hermle pendulum movements also show the pendulum length in centimeters, such as 114cm, measured from the suspension post to the tip of the pendulum. Other Hermle movements were stamped with clock company names such as Howard Miller, Hamilton, and Haid, instead of the Hermle name. This was sometimes followed with the Hermle model number and sometimes not. The main difficulty with the companies using their own marking system, instead of Hermle's, is that it is often difficult to determine exactly which Hermle model and pendulum length you are trying to repair. This may make it harder to order parts or replacement movements, although knowledgeable suppliers can help.

Seth Thomas, in particular, purchased Hermle movements with its own name and numbers stamped on the plate, instead of the name "Hermle". Here is a cross-reference of several Hermle movement numbers and the corresponding Seth Thomas numbers:

Hermle Movement	Seth Thomas Stamp
451-053/85cm	A403-014
451-053/94cm	A403-015
1161-853/114cm	A415-001
1161-053/114cm	A415-002
1161-853/94cm	A415-003

Most older Hermle movements are marked with the date of manufacture, but in 1988 the company started a date coding system with "A". The letter "B" indicated 1989, and so on. Seth Thomas used its own four-digit dating system. The first two digits indicated the year the movement was made; the second two indicated quarter of the year. The first quarter was represented by "03", the second by "06", the third by "10" and the fourth by "13". As an example, a Seth Thomas movement stamped "8203" was made between January 1 and March 31, 1982.

MODELS 340-020 and 1050-020

Hermle spring driven movements are widely used in mantel clocks sold by many different companies. The movements play Westminster or triple chimes. The escapements are based on the pendulum, floating balance, or platform balance. Figure 100 shows the most popular version, Model 340-020, with the floating balance escapement and Westminster chime. The floating balance was discontinued in the late 1980's in favor of a platform balance. Busy repair shops take in many of the 340-020 movements. Many are housed in bracket style cases and were presented as retirement gifts or service awards. They are so common that it is impossible to avoid them. But there is every reason for you to look forward to servicing this model, once you establish a workable procedure and start getting good results.

The subject of cleaning is always controversial. If you disassemble a clock to clean it, you add hours to the job. But ultrasonic machines can, in most instances, clean the movement just as well whether it is assembled or not. The ultrasonic process can get into the smallest spaces between parts quite effectively. It would seem that the ultrasonic is ideally suited to modern movements. After all, a three to five year old movement shouldn't be worn out. It just needs cleaning and oiling to be as good as new.

Unfortunately, most of the modern spring driven movements coming into the shop today cannot be serviced this way. The percentage of clocks coming back within a year could be very high. What is the reason? With the 340-020, 1050-020, and similar models, it is usually pivot and hole wear on the second arbor of the time or strike train. The pivots can be rough enough to wear the holes badly after several years. Less often, one of the chime train pivots on the second, third, or fourth arbors will wear out a hole. The older the movement, the more likely that several holes are worn instead of just one or two. The repair options for the rough pivot and worn hole are to install bushings of bronze or hard brass, then match with a polished pivot, repivoted arbor, or new arbor and wheel.

Problems we see today were created years ago. Since then, manufacturing changes may have improved the quality of various components. But repairers have not helped the situation: many do not routinely disassemble and clean the movements.

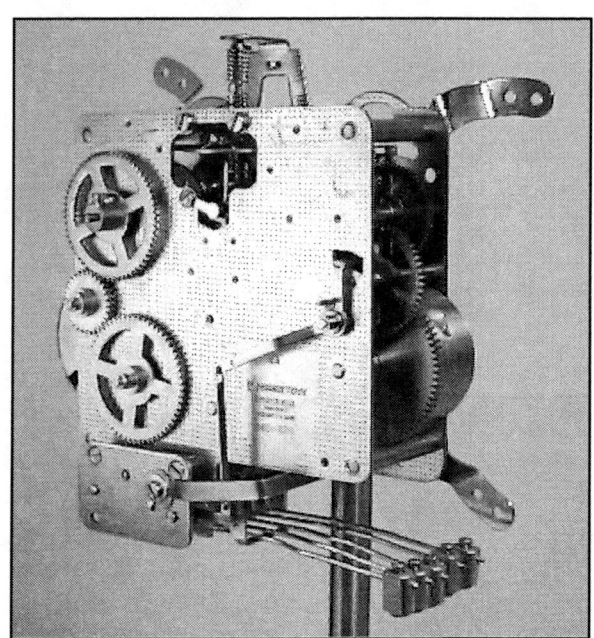

Fig. 100. Hermle Model 340-020 floating balance movement, rear view. This movement is marked "Hamilton" and "81" for 1981.

Instead, they oil dirty movements and hope for the best. Do not avoid disassembling a movement. If it has been running for years, it is likely that it has suffered wear because of dirt and perhaps improper oiling by the customer. However, it can be difficult to do all the work required and purchase replacement parts, without the repair charges exceeding the replacement price of a new movement.

It is very important to clean the three mainsprings. Power from a dirty spring is uneven and much reduced toward the end of the winding period. After a few months of use, the repaired clock may not make it through a seven day run. Inspect the springs for cracked ends, and replace any damaged spring with the exact replacement. Always check the bearing holes on the barrel and cover. If they are worn, don't waste time even removing the old mainspring. Purchase a replacement barrel complete with mainspring, barrel arbor, and cover. Most clock parts suppliers sell Hermle barrels complete with springs. A number stamped on the barrel cover identifies the barrel and spring. The cost, including the barrel, is usually less than a replacement spring for an old clock. You might want to consider removing the new spring from the barrel, cleaning it, then reinstalling and lubricating it. That way you will know the spring is ready to give the best service.

Strike Train Assembly of 340/1050

We'll assume a movement has been cleaned and repaired, and is ready to be put back together. Start by placing all the arbors into the front plate. Leave out the balance assembly until later, to avoid dam-

CHAPTER 14 - HERMLE

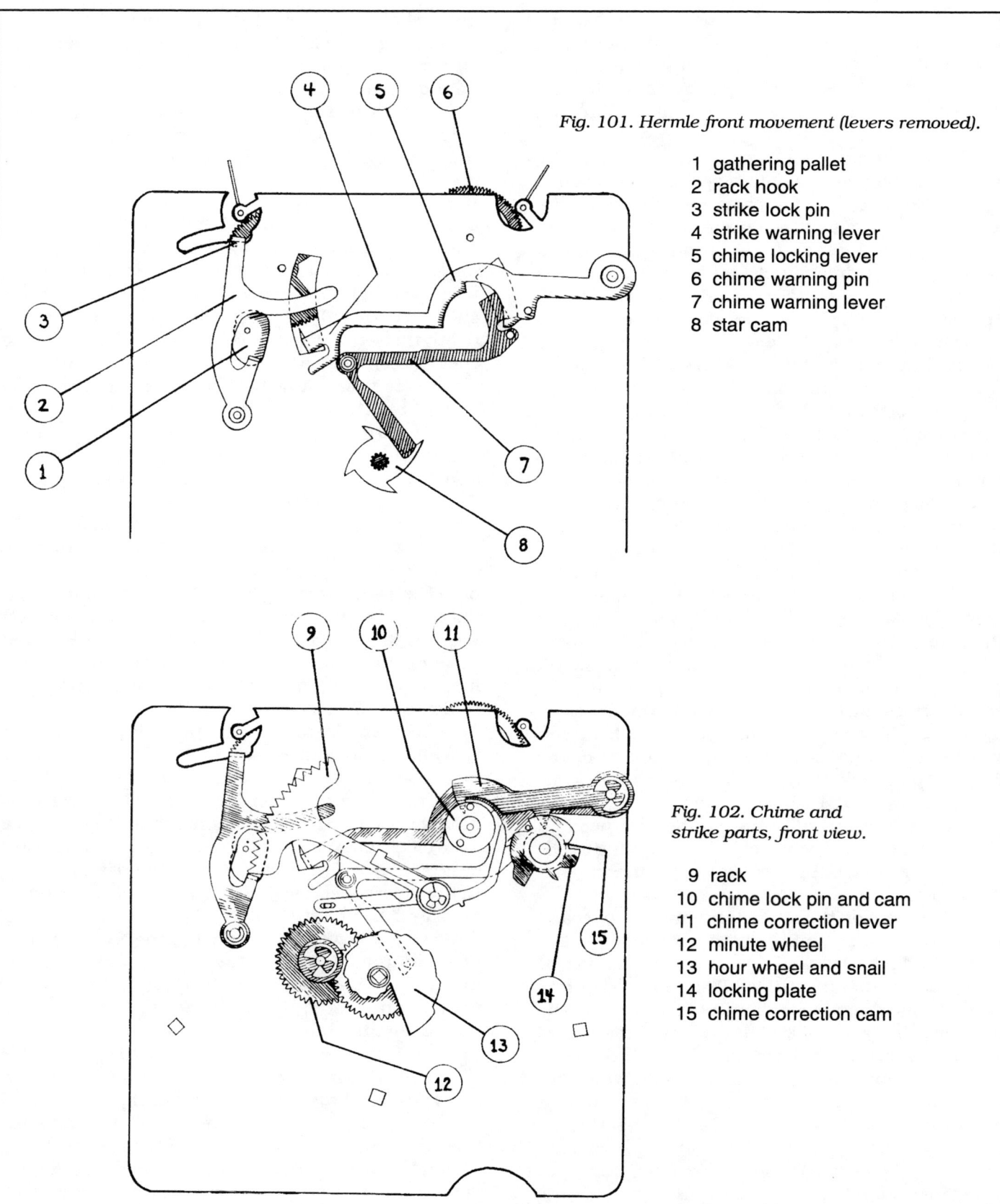

Fig. 101. Hermle front movement (levers removed).

1. gathering pallet
2. rack hook
3. strike lock pin
4. strike warning lever
5. chime locking lever
6. chime warning pin
7. chime warning lever
8. star cam

Fig. 102. Chime and strike parts, front view.

9. rack
10. chime lock pin and cam
11. chime correction lever
12. minute wheel
13. hour wheel and snail
14. locking plate
15. chime correction cam

aging it. Install the rear movement plate. Fit the pivots carefully into their holes, adding the pallet arbor last, and tighten the pillar nuts finger tight. Add the front and rear pivot hole covers for the pallet arbor on the balance movements, after oiling the pivots.

Add the hammer-lift arm on the back of the movement and fasten it with a spring clip. Refer to Figures 101 and 102 for parts identification: the front of the movement is shown in two stages of assembly. Put the rack hook (2) on the post and fasten with a spring clip. Many of the rack hooks have a coil spring for tension. If a straight wire spring is present instead, adjust it for positive action. It is not necessary to bend the spring; it pivots on the post. Add the rack (9) for the purpose of testing the strike train operation. Turn the strike train wheels with your finger. The fly should turn counterclockwise when seen from the front. Run the gear train through a few cycles. Notice the position of the hammer-lift arm on the rear of the movement. When the wheels stop at the end of a cycle, the hammer-lift arm should not be in contact with the hammer-lifting star on the third arbor. If it is, the strike hammers will have to start up under load each hour. To correct the problem, separate the plates enough to allow disengagement of the third wheel from the pinion above. Re-index these and try the gears again for correct rest position of the hammer-lift arm. You can accomplish the same adjustment by turning the gathering pallet (1) to a new location on the elongated pivot. However, the gathering pallet is driven on tight and cannot be twisted unless it is removed first and put back on again. This is more trouble than re-indexing the wheel and pinion. I don't remove the gathering pallet unless the hole needs a bushing, which is rare. I leave the gathering arbor and wheel in the front plate, with the gathering pallet attached.

Another item on the strike train must be checked at this time. Notice that the gathering pallet is also a cam, which raises and lowers the rack hook at every "pass" or hammer blow. The upper portion of the rack hook serves as the locking lever. At the last pass, the rack hook slides beneath the end of the rack, and is finally lowered enough so the lever gets in the way of the strike lock pin (3) which revolves with the locking wheel. Check to be sure that the pin contacts the lever safely. If the gathering pallet is still in the process of lowering the rack hook and locking lever when the pin comes by, the pin may glance off the tip of the lever instead of coming to rest against it. If necessary, re-index the locking wheel pinion with the gathering wheel below it.

One part to check on the strike mechanism is the hammer-lift, part #15-33 on the Model 340-020 movement. On a few of the movements, the hammer-lifting star cuts deeply into the part after years of service. If you find a deep groove, replace the hammer-lift with a new part. Some of these are found badly worn, others hardly worn at all. It may be a matter of dirt and lubrication which causes the variation. Lubricate the new part with light clock grease.

Chime Train Assembly of 340/1050
The plates will not have to be separated again during assembly. Install the three barrels and winding parts, and tighten the pillar nuts. Note that the triple chime 1050-020 model requires a washer underneath the screw which holds the cover for the chime click wheel. This prevents the screw from projecting inside the plate where it would jam the triple chime shift linkage. Wind the chime and strike partially, to provide power for testing the operation of chime and strike together. Referring to Figures 101 and 102, put on the chime lifting and warning lever (7), and adjust the tension spring. You will have to remove the rack (9) first, which was installed to aid in making the strike adjustments. Leave the rack off for later assembly. Next, add the piece that has both the strike warning lever (4) and the chime locking lever (5) on it. Fasten it in place with the setscrewed brass collar that fits over the arbor and rests against the inside surface of the front movement plate. Tighten the set screw after making sure that the arbor has endshake.

Proceed to install the chime lock pin and cam (10) and the locking plate (14). Before tightening up on these two parts, set the chime warning. First, place the warning wheel so that the chime warning pin (6) is in a 1 o'clock position. Figure 101 shows a 12 o'clock position which is also acceptable. Now set the chime lock pin and cam against the notch in the chime locking lever (5). Tighten the set screw to fasten the cam in place. Next, set the locking plate so the pin on the chime locking lever is in one of the four slots. Tighten the set screw to fasten the locking plate. Finally, add the chime correction lever (11) and fasten with a spring clip. You will then have assembled the chime train in the locked position with the correct warning pin setting.

Install the minute wheel (12), hour wheel and snail (13), and rack. Before putting the washer and spring clip over the minute wheel, check the snail position. Do this by turning the minute hand through several quarters, waiting each time for the gear train to function. If the rack tail drops correctly to permit 12 o'clock and 1 o'clock striking, the adjustment is right.

The automatic chime correction mechanism must be checked. It is made up of the chime correction

CHAPTER 14 - HERMLE

lever (11) and a cam (15) on the rear of the locking plate. In Figure 102, the cam and locking plate are shown in position following the third quarter chime. The chime correction lever is lowered into the cam slot, and the end of the chime correction lever is hooked over the chime lock pin. Only at the hour does the chime lifting lever receive enough lift from the star cam (8) to unlock the gear train to chime the hour. Unless the chime correction lever is bent or damaged in some way, no repair will be needed. Just turn the minute hand through several quarters to verify that the mechanism will self-correct.

Next, install the hammer assembly underneath the movement. Hook on the connecting link between the strike hammers and the hammer-lift arm. Add the large chime drive wheel on the elongated rear pivot of the chime third arbor. A small and large wheel are added below to transmit power to the pin barrel.

The chime hammers must be set on the proper sequence. Turn the minute hand through a full hour, until you have completed the four-note first quarter chime. Loosen the set screw on the chime drive wheel, then turn it counterclockwise by hand. This will operate the hammers independently of the gear train. After you see the four chime hammers rise and fall in order from front to rear, the adjustment is done. Tighten the set screw and then check to make sure the hammers never hang up at the end of any sequence. On a triple chime movement such as the 1050-020, watch for all eight hammers to operate in order on the Whittington first quarter chime.

The condition of the chime train in the spring driven movement is best judged by its performance toward the end of the winding period. If chiming is still fast enough, there can't be much wrong with the chime train. With the weight driven movement, we are not concerned over the constant pulling force of the weight. It is either adequate or not, depending on the condition of the movement. But with the spring driven movement, we need to be aware of the diminishing power of the mainspring as the clock runs down. If the chimes are very slow on the seventh day, you are going to have problems with the clock. After a few months, performance will fall off just enough to cause the clock to stop chiming after six days. When you consider that many customers are reluctant to wind the chime mainspring fully, it is even more important to check the running period of the clock. When you set the clock on the shelf for a test run, note the date and time you begin. An eight day run, followed by winding, then another trouble-free running period, is a good test.

One repairer does not oil exactly the same way as another. Each makes a choice of oil and grease, and repairers do not always agree on where it should be applied. As far as the chime trains of the 340-020 and 1050-020 are concerned, there are several places which must be lubricated in addition to the pivot holes. The curved end of each hammer shaft, at the opposite end from the hammer head, is an example. As the hammer falls, the end of the hammer shaft slides into contact with the movement pillar. It establishes the rest position of the hammer and limits bouncing by dampening the shock of the hammer blow. Unless you lubricate the end (or the pillar) with clock grease, the hammer will bind. This one detail can stall the chimes completely if you forget to take care of it.

Another important area is the triple chime shift mechanism on the 1050-020. The selector lever moves the pin barrel forward and back in distinct steps to change chimes. You should grease the steps on the cam so the action will be smooth. The lever should come through the dial far enough to allow a good grip, and should not bind on either side of the slot. If these conditions are met, the chime selector will work smoothly and give the owner the feel of moving the lever in steps. If the lever binds, he may accidentally push it between two selections. This will produce strange sounds or jam the chime train.

Finish by adding the balance unit (or pallets in other models). Most of these movements are fastened in the case by means of four removable wing-style brackets. It can be frustrating to get the movement placed correctly behind the dial and still get all four nuts tight, with the brackets over the screw holes. A simple procedure will make the job easier. Before disassembling the movement, mark the brackets so you will be able to put each one back on the same pillar. Now it is just a matter of positioning the brackets over the screw holes and tightening. The movement will be correctly placed. Always bear in mind that the nuts may come loose during your attempts to adjust the brackets, and that you are holding a spring driven movement with "power on".

MODELS 451/1151

Hermle mechanisms are the most popular and widely used movements today. Hermle retains the same basic movement design, constantly modifying and updating the product. As for the repairer, he or she is often placed in the same "squeeze" as with Hermle movements from relatively inexpensive mantel clocks. The 451/1151 series movements have been extensively used in low-priced floor clocks and kit grandfather clocks. Customers often balk at paying for the kind of work which is required. Replacement movements will be easier to sell than a repair in some instances.

Many of the clocks have already had the "easy" servicing done, the first one or two oilings. Now, after five to ten years of use, they will need more expert attention. The problems you can identify at this stage will, in some cases, be latent defects. The 1977-1979 manufacturing years left a legacy of soft leaded steel pivots which did not hold up. The resulting problems were worsened in most cases by neglect (no servicing at all) or abuse (improper servicing). After 1979 improvements were made which have presumably lengthened the useful life of the movements and made repairs (bushing work, in particular) more likely to last.

Most of the movements you encounter will need cleaning. If no manufacturing defects are present, the customers usually call for service after dirt and lack of lubrication cause the movement to fail. You may clean ultrasonically, but virtually every movement should be taken apart, no matter how you clean it. There will probably be one or two worn holes, and there may be evidence of black deposits in them. After disassembly, you will find that the corresponding pivot is rough. Putting the rough pivot into a new bushing is the same as doing nothing at all; rapid wear will occur.

A number of repair options exist, as with the mantel movements. New arbors and wheels are available. Bronze bushings have given good service, and brass bushings of increased hardness have also been offered to the repairer. It is just as important that the pivot be made smooth. Unless the roughness is polished out or the entire wheel and arbor replaced or repivoted, your repair will not last.

Strike Train Assembly of 451/1151

When you oil or restore a five to ten year old Hermle movement, you must test the results of your work. The "test" is either a few minutes of observation following home servicing, or a week or two on the test stand in your shop after a restoration. In either case, you are looking for the same things.

If the strike train runs, you cannot stop there. You must look for any signs that will predict trouble on the way. Look for any weak spot that is going to stop the clock. One advance warning is a worn part. Leaving it alone is unwise, especially when replacement parts are available. You're asking for a failure at some unpredictable time. Another tip-off of trouble ahead is poor performance during the test. If you see something which isn't right, you can count on it getting worse later on.

Let's begin by looking for worn parts. Hermle weight driven movements made in the 1970's often show wear in the hammer tail. Figure 103a shows the old style hammer arbor and hammer tail, which Hermle calls the hammer-lift, part No. 16-124. Over the years, the hammer-lifting star may have cut a

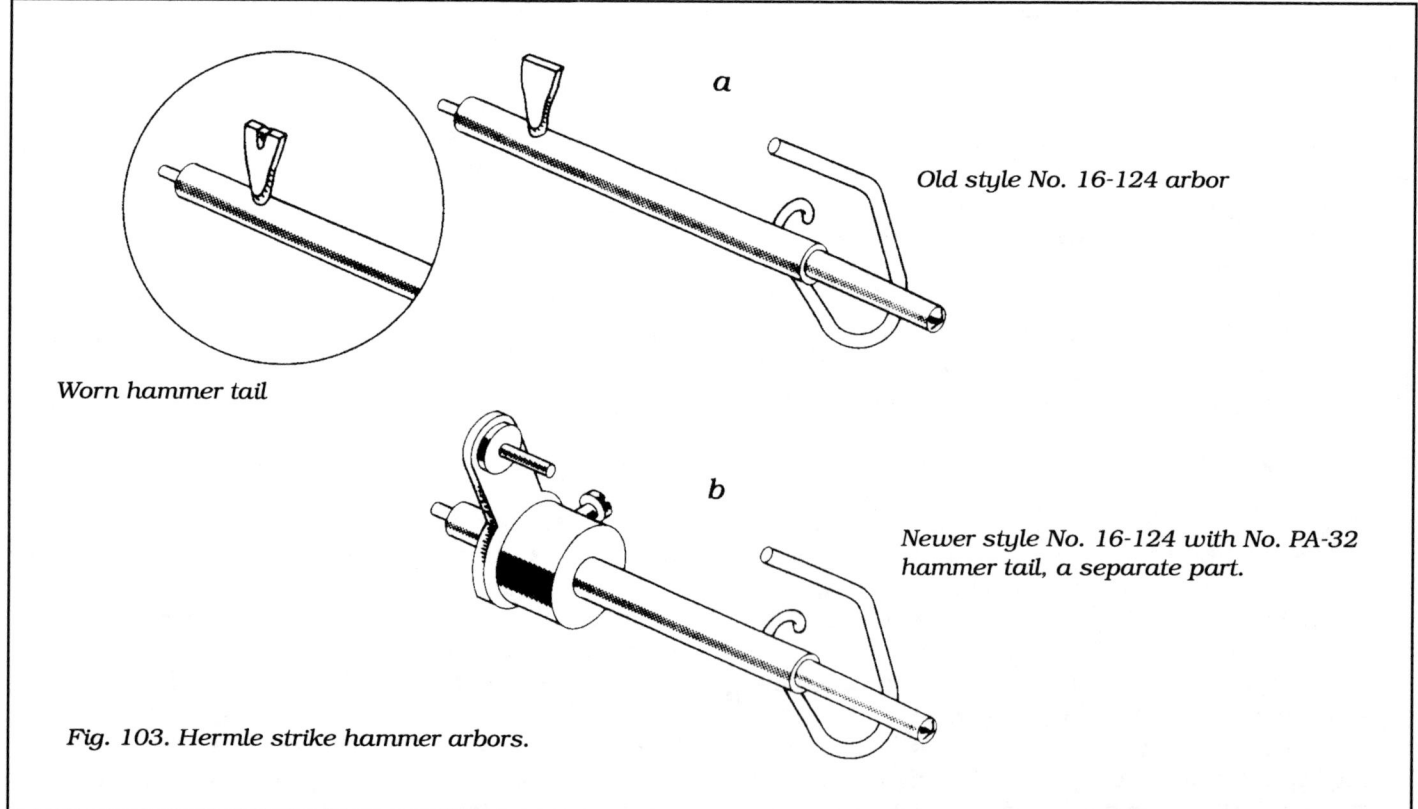

Worn hammer tail

Old style No. 16-124 arbor

Newer style No. 16-124 with No. PA-32 hammer tail, a separate part.

Fig. 103. Hermle strike hammer arbors.

CHAPTER 14 - HERMLE

deep groove in the hammer tail, as shown in the inset. It's a good idea for you to replace the worn part.

The piece used in newer movements is shown in Figure 103b. It consists of two parts. The newer No. 16-124 is the same as the old part, but without the hammer tail. A separate hammer tail, part PA-32, is added. The new hammer tail is a polished steel pin. To make the change, you must take the movement apart. Cut off the old hammer tail and file the arbor smooth on both sides. Slip the brass collar and steel pin combination, No. PA-32, onto the arbor. Fit the assembly between the plates, along with the wheel which carries the strike hammer-lifting star. Attach the hammer spring, shown in Figure 104. Adjust the position of the hammer tail until it works freely and without interference. Then tighten the set screw. This procedure will take only a few minutes, and it's worth it to prevent a failure later on.

Observe the strike train in operation. First, make sure the strike speed is fast enough. Experience is the best teacher. A slow-striking clock will soon slow down even more, then fail. Next, watch the fly as the gear train runs, to detect momentary changes in the speed of the wheels. In some movements, the strike train slows almost to a stop each time the hammer-lifting star encounters the load of the hammer tail. The load threatens to stall the gear train with each hammer blow. If the speed picks right up after the hammer falls, you might not even notice the problem unless you look for it. Check for oil first. During an in-home servicing, you might have forgotten to add oil to one of the strike fly pivot holes. If a dry hole was causing slow striking, then all the oil you add to the other pivots will not solve it. Next, consider whether general dirt and wear are worse than you thought. You may need to do a cleaning after all.

Fig. 104. Strike hammer spring (shown with old style hammer arbor for clarity).

Another possibility exists. The hammer spring, shown in Figure 104, may be too strong. I've replaced this spring in several movements. It is hard to determine whether the original spring is correct. But you can try installing a slightly weaker spring. Observe the strike train again. If the fly doesn't slow down with each hammer blow, then use the new spring. Remember, however, that a weaker spring should not be used to compensate for other problems. The spring is there to push the hammers against the chime rods. A very weak spring cannot do the job, and the strike will sound weak. The correct spring for the 451/1151 movements is number M-190C. The larger movements 461/1161 use spring number M-190B. Although the M-190B is stronger, the difference is slight and it is hard to tell the two springs apart by looking at them.

Chime Train Assembly of 451/1151

The speed of the gears is the best indicator of the condition of the chime train, just as it is with the strike. Any movement that chimes too slowly will just get worse. Some of the older movements stall, and others chime slower and slower, until the hour chime seems to take an eternity to finish. If oiling brings the speed back to normal, you have probably done all that you can do during a service call in the customer's home.

Just as the strike has a critical load point each

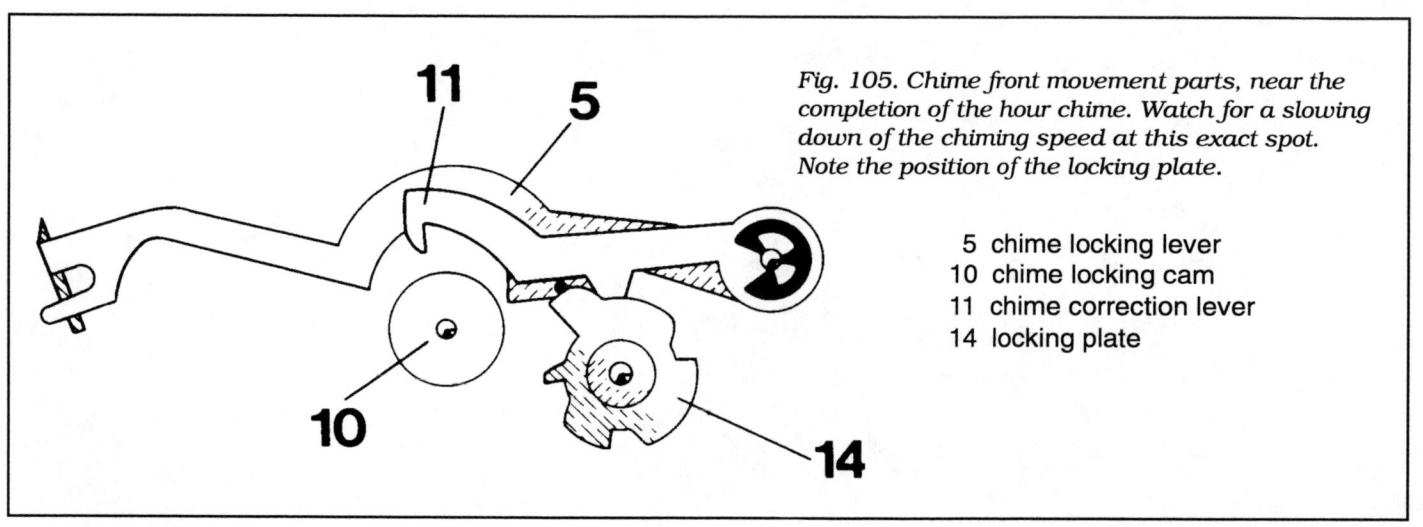

Fig. 105. Chime front movement parts, near the completion of the hour chime. Watch for a slowing down of the chiming speed at this exact spot. Note the position of the locking plate.

5 chime locking lever
10 chime locking cam
11 chime correction lever
14 locking plate

time the hammers are lifted, the chime train has its own critical time. During the hour chime, the long lever across the front of the movement raises the rack hook to send the strike train to the warning position. The added load will not slow the chime speed if the clock is in good condition. But a worn, dirty chime mechanism will slow down, and may nearly stop, each hour. Figure 105 shows the exact moment when the problem will show up. Note the locking plate (14), which is lifting the lever (5) to the highest point. By observing the clock in operation with the dial off, you can see in a moment whether the problem exists. If you cannot do anything to eliminate it in the customer's home, then you should recommend shop work.

15

URGOS 06-SERIES

Urgos, the German manufacturer, produced its 06-series movements in at least five variations on a basic spring driven movement with plates 110mm square. Overall movement height is actually 128mm because of the hammer assembly. Model UW 06061A was studied for this chapter. The back plate is marked UW 6/61A. (UW stands for *Uhrwerk*, or clock movement.) This one is fitted in a Barwick, (Howard Miller) wall clock with empty weight shells to give the appearance of a weight driven clock. It plays Westminster chimes and strikes the hours on five bottom-mounted hammers.

DISASSEMBLY

To restore the movement, take it completely apart. First remove the suspension spring and pendulum leader to avoid damaging them. Next, let down the mainsprings completely. Then remove the click wheels and their covers. Pull the winding arbors straight out through the front of the movement. Notice that a squared portion of each arbor fits into a corresponding shape inside the barrel arbor. It is only the click wheel and cover which hold the arbor in the clock. Remove the chime and time barrels first, and the strike barrel last. The strike barrel must come out through the bottom center of the movement rather than the strike side.

The three winding arbors are the same, as are the click wheels. The time and strike mainsprings are the same: Urgos specified them as 17mm wide, 0.38mm thick, and 1300mm long. The chime spring is 24mm by 0.42mm by 1500mm. Replace any worn or damaged winding parts.

Continue with disassembly. Referring to Figures 107 and 108, take off the rack (24), hour wheel (23), minute wheel (20), silencer lever (18), and rack hook (1). Loosen the set screw (13) and remove the chime-strike lever (6) through the front. The chime locking lever (9) and chime correction lever (10) are mounted on the arbor for the chime-strike lever, and can be removed when the arbor is pulled out. On this movement there is a thin brass washer between parts (9) and (10). There is also another part to keep track of, a coiled tension spring acting on the chime strike lever. Now remove the chime lifting lever (19), noting the fact that there is a smaller coiled tension spring under it.

After taking off the locking plate (16), you will find the front of the movement clear except for the gathering pallet (4). Leave this attached to the locking wheel unless there is some compelling reason, such as a worn pivot hole, for removing it.

From the rear of the movement, take out the pallet bridge and pallets, the large chime drive wheel, the large and small idler wheels under it, and the pin barrel wheel on the bottom. Remove the strike hammer-lift arm and lifting wire. Now you are down to the bare movement.

Before separating the plates, note the four movement brackets, one located at each movement pillar, under the nut. The brackets swivel unless tight-

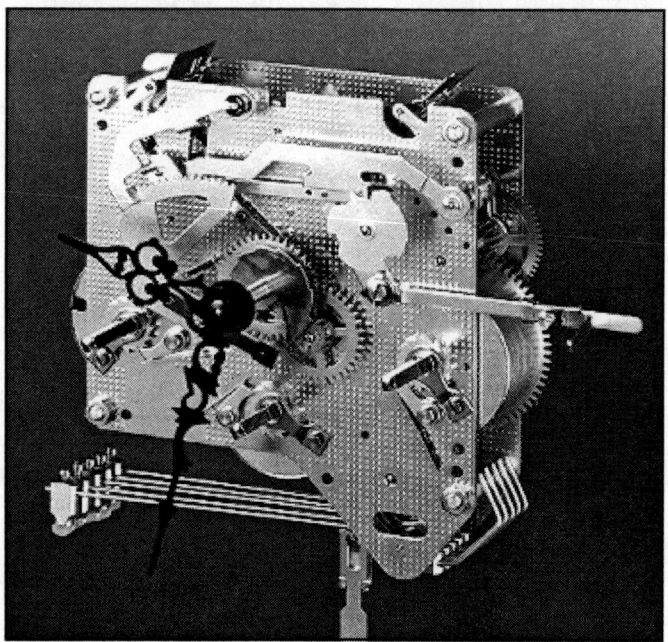

Fig. 106. Urgos 06-series. Photo courtesy of Urgos.

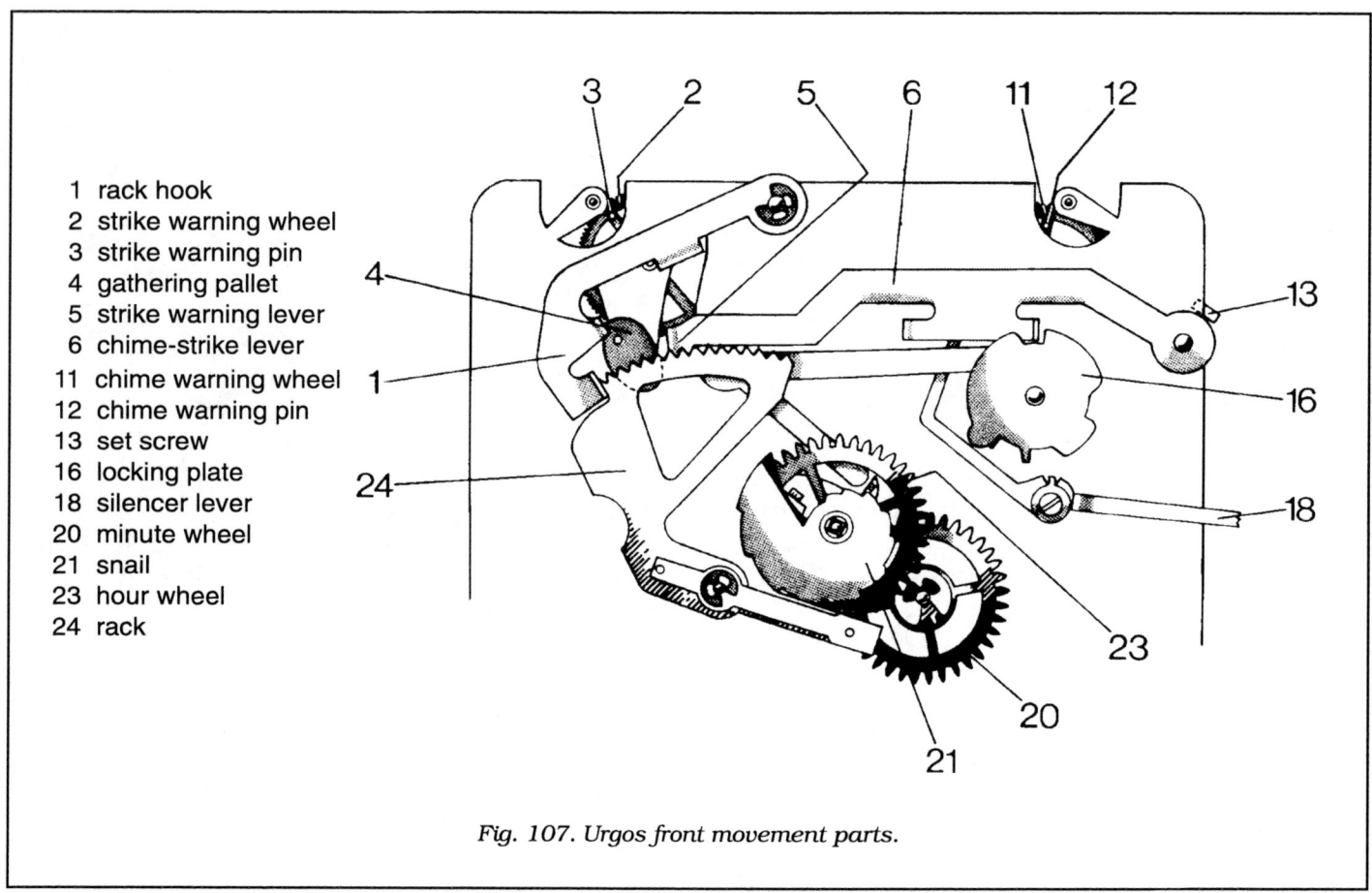

Fig. 107. Urgos front movement parts.

1 rack hook
2 strike warning wheel
3 strike warning pin
4 gathering pallet
5 strike warning lever
6 chime-strike lever
11 chime warning wheel
12 chime warning pin
13 set screw
16 locking plate
18 silencer lever
20 minute wheel
21 snail
23 hour wheel
24 rack

ened under the pillar nuts, and they serve to locate and fasten the movement in the case. It is far easier to reposition the movement after repair if the brackets are not switched to different pillars. Mark them upper right, lower right, and so on. Remove them and reinstall the nuts. Then remove the rear pillar nuts and take off the back plate. The pillars can stay attached to the front plate, without the brackets being there to get in the way. All the arbors will go back into the front plate first, during assembly. The movement will be easier to assemble this way, because the center arbor and strike locking wheel will stay with the front plate.

The only wheels one might want to mark for identification are the second wheels in the time and strike trains. Remove the wheels, pin barrel, hammers, hammer shaft, and the lower strike hammer-lift arm. Mark the hammers so you can get them back on the shaft in the right order. Polish pivots and install bushings, as necessary, and clean the movement.

REASSEMBLY AND ADJUSTMENT

Load all the arbors into the front plate. Install the pin barrel with the long pivot pointing toward the rear of the movement. The hammer shaft goes in with the brass collar toward the front. Add the five hammers and the lower hammer-lift arm. Put on the back plate and fit all the pivots into the holes. Some of the pivots are fine, especially the fly pivots, so be careful not to bend them. Screw on the pillar nuts finger tight only.

Strike Adjustments

Install the rack hook (1) on the front of the movement, and add the upper hammer-lift arm, which engages the hammer-lifting star on the third wheel. Rotate the wheels with finger pressure, with the strike fly moving counterclockwise. If you unlock the strike train, it will strike once and lock again. Locking is accomplished on the front of the movement, as the rack hook comes to rest against the notch in the gathering pallet (4). When the train locks, the hammer-lift arm must be clear of the hammer-lifting star. This is necessary to avoid hammers left in the raised position at the end of the hour strike. To adjust, separate the plates enough to change the mesh of the third wheel and the next higher pinion.

Before going further, install the chime-strike lever (6) and its tension spring. Place the chime correction lever (10) and chime correction arm (14) in position before inserting the arbor through the

CHAPTER 15 - URGOS 06-SERIES

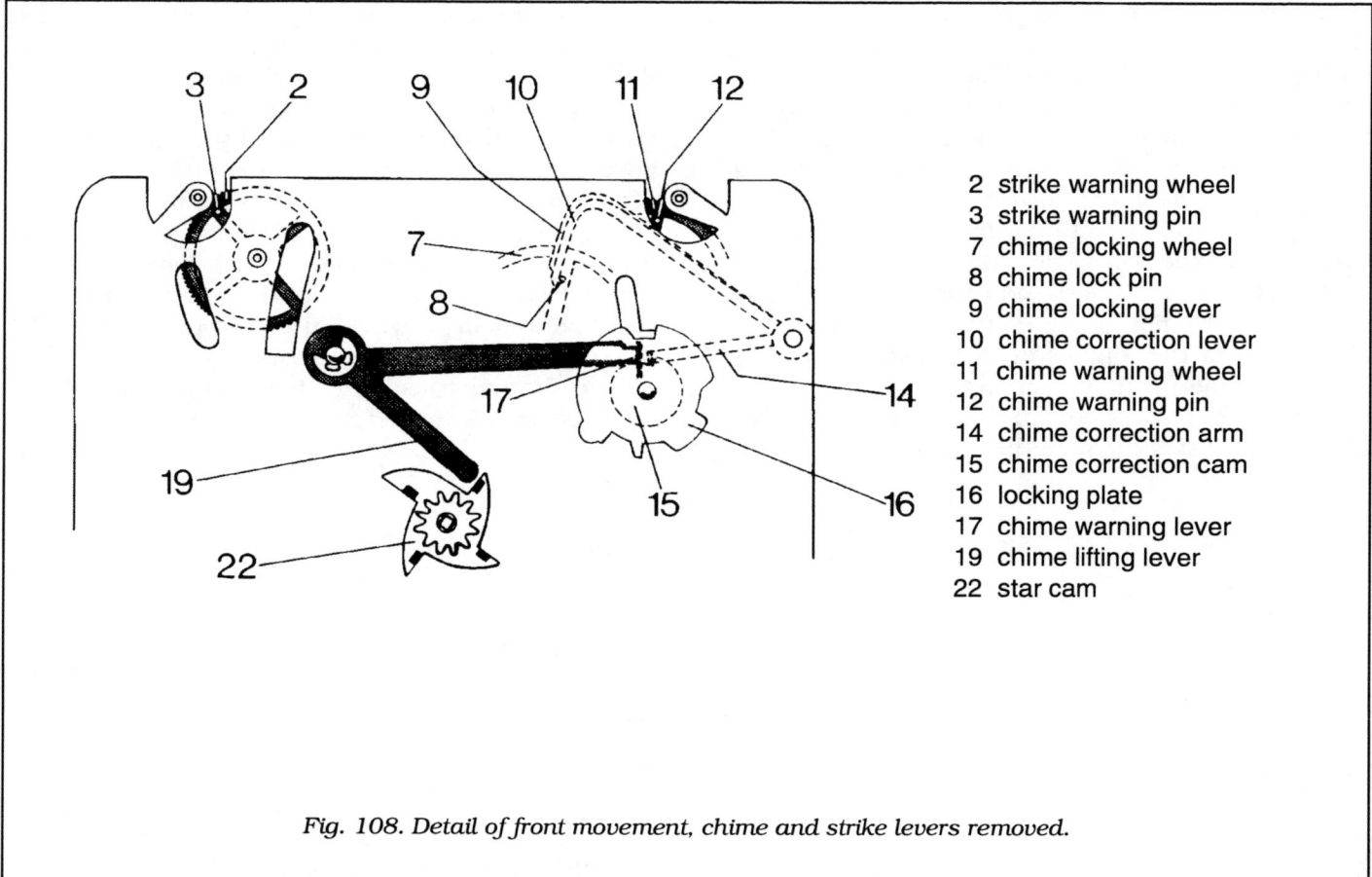

2 strike warning wheel
3 strike warning pin
7 chime locking wheel
8 chime lock pin
9 chime locking lever
10 chime correction lever
11 chime warning wheel
12 chime warning pin
14 chime correction arm
15 chime correction cam
16 locking plate
17 chime warning lever
19 chime lifting lever
22 star cam

Fig. 108. Detail of front movement, chime and strike levers removed.

plates. The chime correction arm goes in front of the chime warning wheel (11), and the chime correction lever is located behind the same wheel. The chime correction lever must have access to the chime lock pin (8). Start the long arbor through. Now add the chime locking lever (9) behind the chime correction lever. The two parts are the same shape, and they operate right next to each other. Notice the small tab on the underside of the chime locking lever. Its function is to lift the chime correction lever each quarter hour. Insert the arbor all the way through to the back plate. Tighten the set screw (13) lightly.

Now that the chime-strike lever is in place, check the strike warning run. The strike warning lever (5) is at the extreme left end of the chime strike lever. When the strike train is locked, the strike warning pin (3) should be at a 12 o'clock position. This will allow almost a half revolution of the strike warning wheel (2). To adjust the warning, separate the plates just enough to permit disengaging and moving the warning wheel to the desired spot while the locking wheel remains still.

Chime Adjustments

The chime warning is checked in the same manner. Install the chime lifting lever (19), which has the chime warning lever (17) at its end. Rotate the wheels by hand, with the fly turning counterclockwise, until the chime locking lever stops the chime lock pin. The chime warning pin (12) should be at approximately a 12 o'clock position to allow for a half revolution of the chime warning wheel, before the pin contacts the warning lever. If necessary, separate the plates slightly and correct the mesh of the locking wheel and the pinion above.

With the warning action properly set for the chime and strike trains, install the mainspring barrels before finishing up the assembly and adjustment procedure. After the barrels and winding parts are in place, put the four movement mounting brackets back on the movement pillars. Tighten all the pillar nuts, front and back. Wind the chime and strike springs partially. Do not wind the time spring until the pallets are in place.

Setting the depth of the chime locking lever is critical. If it locks too deeply, you may have no chiming at all. A shallow lock may mean the train fails to stop when it should. To check the locking action, install the locking plate and then turn the minute hand through several quarters. Adjust the position of the chime locking lever, and then tighten the set screw. Verify that the chime correction feature operates. Remember to leave some endshake on the arbor, which will assure freedom of motion for the chime-strike lever.

The chime correction mechanism is simple in design, and does not require any adjustment as long as the chime locking lever is working properly. The locking plate (16) has a chime correction cam (15) mounted behind it. Following the third quarter chime, the chime correction arm (14) drops into the single slot in the cam. The chime correction lever (10) is lowered into the path of the chime lock pin (8). No further chiming occurs until the hour, when the longest of the four arms of the star cam (22) lifts the chime locking lever high enough so that the tab on its underside can lift the chime correction lever out of the way of the chime lock pin.

To complete the assembly, install the rack (24), then the snail and hour wheel (21 and 23), and the minute wheel (20). Adjust the position of the snail so that 12 and 1 o'clock are struck correctly, before adding the large split washer on the minute wheel post. On the rear of the movement, put on the chime drive wheel and the three others below it.

Adjust the chime note sequences by chiming the clock through the first quarter and then setting the hammer action to agree. Tightening the set screws on the chime drive wheel and pin barrel wheel will firm up the adjustment. On this movement, the first quarter hammer sequence is four hammers rising in order from front to rear.

16
URGOS 9-TUBULAR BELL

In this chapter we will study the assembly and adjustment of the 9-tubular bell movement by Urgos of Germany. First available in the fall of 1976, it was installed in "top-of-the-line" clocks by American clock companies. Until the Urgos appeared on the scene, Herschede produced the only 9-tube movement for the American market. However, Herschede stopped production in 1984. Later, German makers Hermle and Kieninger began producing 9-tubular bell movements. Urgos stopped manufacturing in the early 90's, but the company reorganized and started up again in 1994, mainly supplying replacement movements similar to the earlier models.

Figure 109 is an overall view of the Urgos movement mounted on the seatboard, with tubes, weights, and pendulum installed. Fortunately, the basic movement has much in common with other Urgos movements. Other models in the 03 series were made in large numbers, especially the chain driven Westminster and triple chime rod movements. These have the same basic pattern of chime and strike levers. Similarities in design with the 9-tube movement are visible at once. I am not saying that all these mechanisms are the same, for they are not. The greatest difference is in the addition of the tubular triple chime mechanism and hammer assembly in place of the rod chime configuration.

Fig. 109. Urgos 9-tube movement. Photo courtesy Urgos and Carl J. Bendorf, manufacturer's representative.

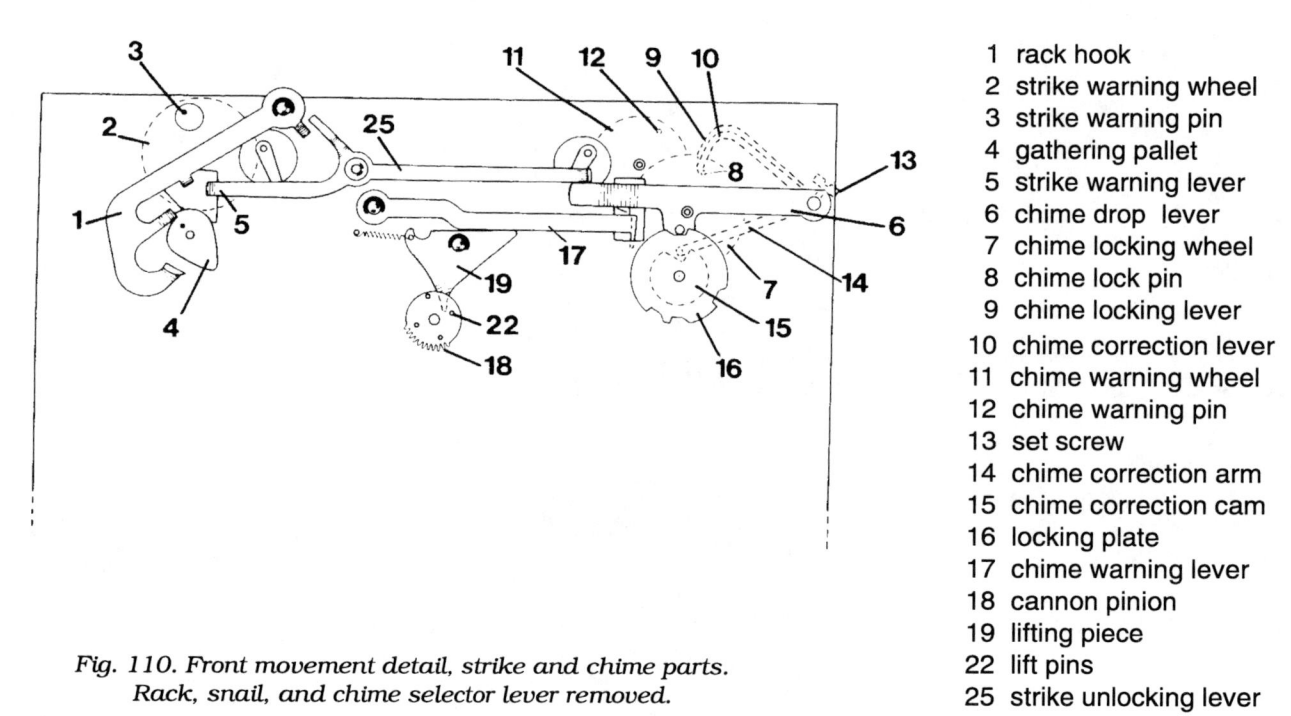

Fig. 110. Front movement detail, strike and chime parts. Rack, snail, and chime selector lever removed.

1 rack hook
2 strike warning wheel
3 strike warning pin
4 gathering pallet
5 strike warning lever
6 chime drop lever
7 chime locking wheel
8 chime lock pin
9 chime locking lever
10 chime correction lever
11 chime warning wheel
12 chime warning pin
13 set screw
14 chime correction arm
15 chime correction cam
16 locking plate
17 chime warning lever
18 cannon pinion
19 lifting piece
22 lift pins
25 strike unlocking lever

Urgos technical sheets list the time and strike weights as 3.5 kg each, and the chime weight as 8 kg. Urgos made the movement in several different pendulum lengths. Some models have a wooden pendulum shaft, and others are the popular lyre style. The chimes are Westminster, Whittington, and St. Michael. The owner selects the desired chime by moving the chime selector lever at the numeral "3". If the chimes are placed on "silent", the hour strike is also silenced. All nine tubes hang in a row, with the longest (for the hour) on the left.

DISASSEMBLY AND REPAIR

Refer to Figure 110, front detail view, and Figure 111, complete front view, as you proceed. We'll assume the weights have been removed and the movement has been taken from the clock cabinet. Begin by unhooking the hammer cords and taking the movement off the seatboard. Remove the entire chime drive mechanism (35), and the pin barrel (31) and hammer assemblies (27 and 30) as a unit. Set these aside for cleaning.

Before you go further, some care is necessary with the second hand arbor. In this movement, the escape wheel rotates counterclockwise. It was therefore not possible to simply lengthen the arbor and install a second hand. To accommodate a second hand, a mechanism was devised. A seconds transmission wheel (38) is located on the end of the escape arbor. It drives a wheel on the end of the seconds pivot (37), producing the correct rotation and speed for the second hand. This mechanism is delicate, so you must be careful with it. First, remove the seconds front plate (36) and the seconds pivot which it supports. Now you will notice that the seconds transmission wheel remains on the end of the escape arbor. You must not pry off this small wheel. You will almost certainly nick a tooth or bend the escape pivot, rendering the escapement inoperable. Later, after the entire movement is apart, you can support the front plate and safely drive the escape arbor through (down) to remove the seconds transmission wheel.

Proceed to separate the plates and remove all wheels. If the pivot hole is not worn, you may want to leave the gathering pallet (4) in place instead of driving it off. Remove the bracket for the seconds pivot, fastened with two small hex nuts behind the front movement plate. Clean the movement, and inspect for worn holes or rough pivots. In particular, check the hole behind the minute wheel (20). The tension spring (43) provides hand tension. Sometimes the force of the driving weight accelerates hole wear at this location, and the minute wheel

CHAPTER 16 - URGOS 9-TUBULAR BELL

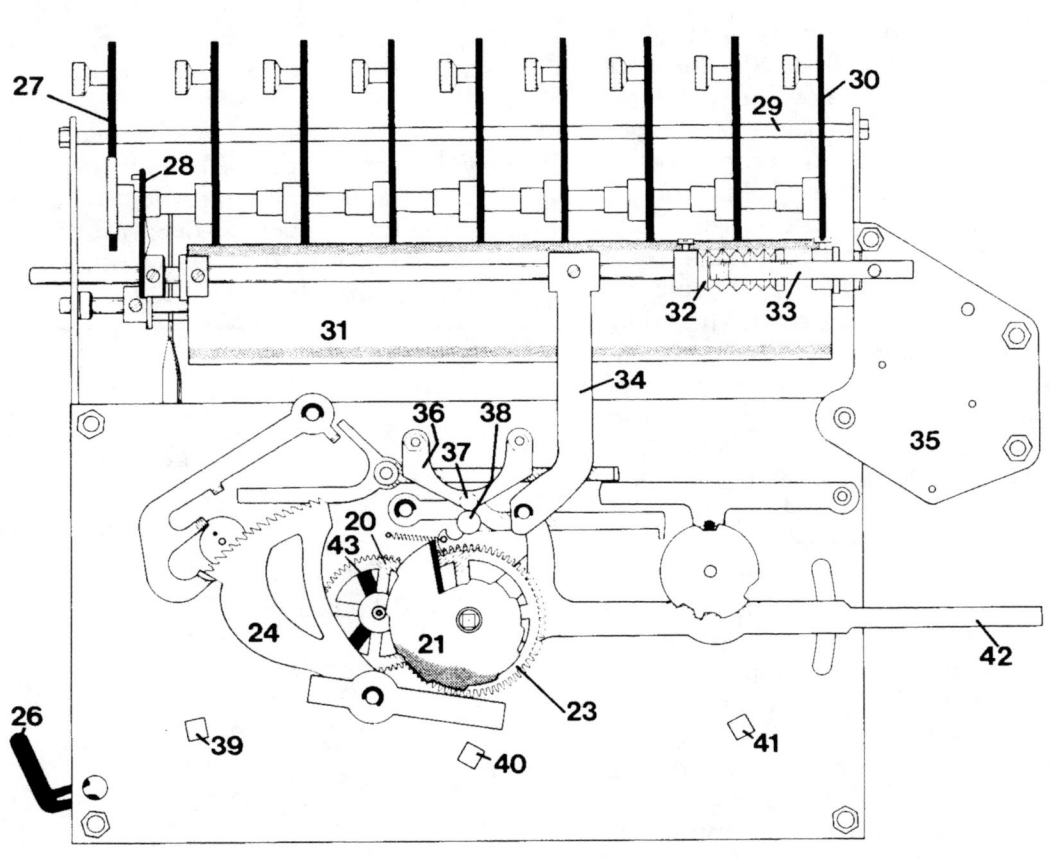

20 minute wheel
21 snail
22 lift pins (see Fig. 110)
23 hour wheel
24 rack
25 strike unlocking lever (Fig. 110)
26 dial clip (one at each pillar)
27 strike hammer
28 strike silencer
29 buffer
30 chime hammers (8)
31 pin barrel

32 chime silencer cam
33 chime selector spring
34 chime shift lever
35 chime drive mechanism
36 seconds front plate
37 seconds pivot
38 seconds transmission wheel
39 strike winding arbor
40 time winding arbor
41 chime winding arbor
42 chime selector lever
43 tension spring

Fig. 111. Urgos 9-tube movement, front plate, complete view showing major parts.

teeth can butt against the cannon pinion (18). This is enough to stop the movement. Polish the pivot and install a bushing if necessary.

REASSEMBLY

One of the first things to do before reassembling the clock is to look for the dial clips (26). The four clips must be pushed onto the movement pillars. Next, install the seconds transmission wheel (38) back onto the escape arbor. To accomplish this safely, insert the escape pivot through its hole in the front plate, then push on the wheel lightly. Lay the entire front plate face down on your anvil, resting it on the wheel. Use a hollow punch placed against the escape pinion, gently driving the arbor through the hole in the seconds transmission wheel.

Add the chime warning lever (17) to the front of the movement now, because the second hand mechanism will go over it. Install the back plate and bracket for the seconds pivot (37), fastening with small hex nuts and washers behind the front movement plate. Add the seconds pivot and its small wheel, followed by the seconds front plate (36). It is essential for you to carefully adjust the depth of the small wheel with the seconds transmission wheel (38). You'll notice that the bracket mounting hole on the strike side of the front movement plate is elongated to permit adjustment. Check the smooth running of the second hand mechanism by putting the movement plates together and spinning the escape wheel alone. If the rest of the time train is left aside, you can easily detect the slightest binding of the two small wheels. When you are satisfied, tighten the mounting nuts.

Now install all the arbors in the front plate, then add the back plate. Fit all the pivots into their holes. One point of confusion may present itself in the chime train. Until the chime drive mechanism (35) is added later, the gearing of the chime train is not continuous. There is a gap. Don't worry about it, because it will be filled in when you install the mechanism. We are going to cover the chime drive setup later in this chapter.

Adjust the strike train first. Add the rack hook (1). Turn the wheels until the notch in the gathering pallet (4) locks in place. Now see if the hammer tail is clear of the hammer-lifting star. If it is not, separate the plates and correct it. Set the strike warning pin (3), as shown in Figure 110, to allow for the warning action, and you are done.

Chime adjustments are next. Run the chime wheels until the chime lock pin (8) rests against the chime locking lever (9). The chime warning pin (12) should be in a 2 o'clock orientation to allow for warning action. Basic chime adjustments are complete. Add the locking plate (16) and set it so the chime drop lever (6) rests in one of the four slots when the chime train is locked at rest.

Take some time to set the depth of the chime locking lever (9). Check to verify that it stops the chime lock pin (8) as you rotate the gears through the quarter hour chimes. If the chime correction feature also operates, then the adjustment is correct. Tighten the set screw (13). If the chime train fails to lock at the end of each sequence, however, you will know the chime locking lever is set too high to catch the lock pin. On the other hand, you will know the lever is set too deeply if the clock will not chime at all. Add the rest of the front movement parts illustrated in Figure 111.

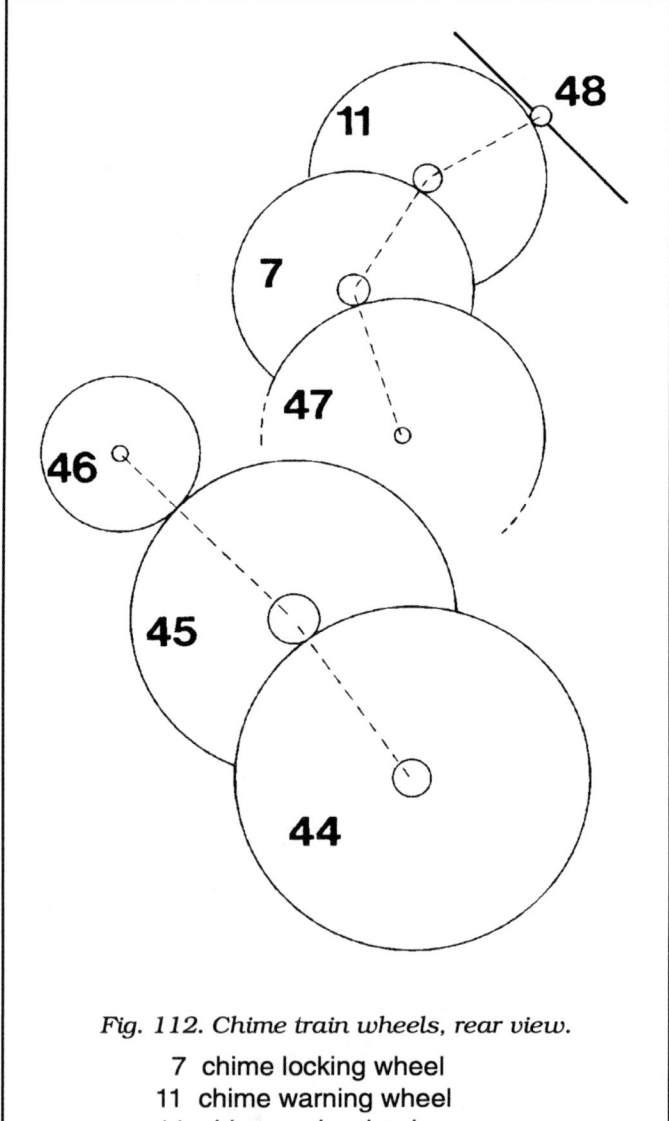

Fig. 112. Chime train wheels, rear view.

   7 chime locking wheel
11 chime warning wheel
44 chime main wheel
45 chime second wheel
46 intermediate chime wheel I
47 chime third wheel
48 chime fly

CHAPTER 16 - URGOS 9-TUBULAR BELL

CHIME TRAIN ASSEMBLY

If you are in the midst of reassembling the movement, you might be having difficulty getting all the wheels in the right places before adding the back plate. The chime wheels don't form a continuous train until you add the chime drive mechanism later on. So you have to put the chime portion of the movement in place with a gap in it. We'll begin by tracing the path of the wheels, to try to remove some of the confusion.

Figure 112 is a diagram of the chime train as seen from the rear, before the back plate is added. The chime drive mechanism is not yet in place. If the chime wheels are correctly installed, you will see two groups of three wheels each. The groups are not connected. Our objective is to install them, so we can put the plates together. We can then proceed to add the chime drive mechanism.

The chime main wheel (44) is the first step. It matches up with the second wheel (45) through its pinion. Power then moves to the intermediate chime wheel I (46). This is a plain arbor with wheel, but no pinion. It transfers rotation out of the regular train, to the chime drive mechanism. We haven't installed that mechanism yet, so the intermediate chime wheel I is a dead end at this point. It is easy to identify because of the lack of a pinion.

Now we go to the other side of the "gap" in the gear train, picking up the trail with the chime third wheel (47). It is also without pinion, but is easy to spot; front and rear pivots are elongated. The front is for the locking plate; the rear is for the chime drive wheel. The lack of a pinion is curious at first—how is the wheel to receive power without one? Actually, the third wheel gets its power from the chime drive mechanism, through the chime drive wheel, on the rear of the movement.

The rest of the chime train consists of the locking wheel (7), warning wheel (11), and fly (48). Now the two groups of chime wheels are in place between the movement plates. The chime drive mechanism will fill in the gap between them.

CHIME DRIVE MECHANISM

It's easy to become confused over the placement of parts in the chime drive mechanism. The mechanism is added after the pin barrel (31) is installed. It is a box-like affair made up of wheels and arbors between two plates, located at the upper right corner of the movement. Its purpose is to transfer rotation from the chime train to the pin barrel. I have used the Urgos designation "intermediate chime wheel" to describe several of the wheels in the chime drive mechanism, because the term provides a good description of what they do. They simply transfer motion.

Figure 113 shows the layout of the chime drive mechanism as seen from the rear of the movement. Install the front plate for the mechanism first, then add the wheels, and finally put on the rear plate for the assembly. The diagram traces the flow of motion beginning at the intermediate chime wheel I (46). This wheel is part of the regular chime train, you'll recall. It transfers power out of the chime train,

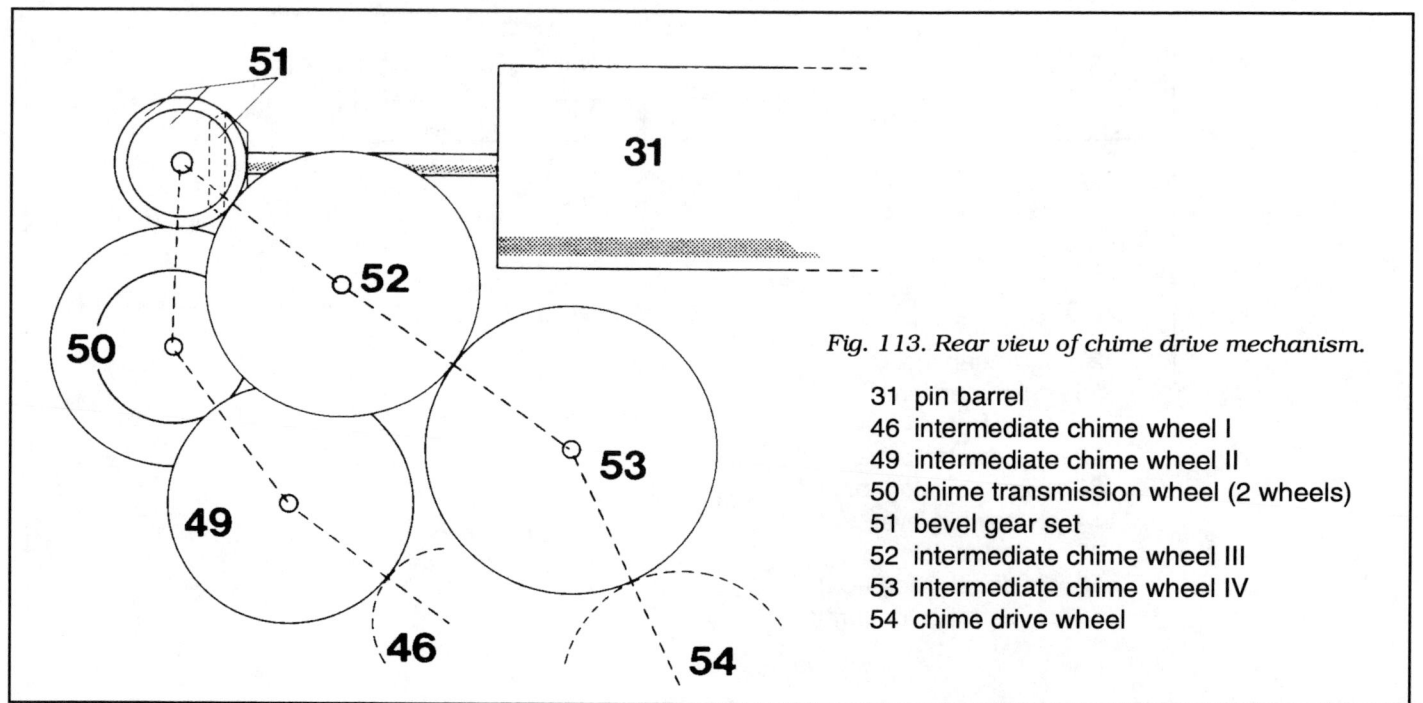

Fig. 113. Rear view of chime drive mechanism.

31 pin barrel
46 intermediate chime wheel I
49 intermediate chime wheel II
50 chime transmission wheel (2 wheels)
51 bevel gear set
52 intermediate chime wheel III
53 intermediate chime wheel IV
54 chime drive wheel

to the chime drive mechanism. The next link in the series is the intermediate chime wheel II (49), a plain arbor with wheel. Next is the part Urgos calls the chime transmission wheel (50), which is actually two wheels mounted on an arbor. Power flows to the smaller wheel, then from the larger one to the bevel gear set (51).

The bevel gear set is a critical group which changes the motion of the wheels to the front-to-back rotation of the pin barrel (31). The bevel gear on the end of the pin barrel arbor receives power from one member of the gear set, then transfers it to an identical member of the group. Having achieved its purpose of driving the pin barrel arbor, the chime drive mechanism now must transfer power back to the rest of the chime train. Since the chime train is interrupted as shown in Figure 112, there must be a transfer of power back into it. We still need to run the rest of the chime wheels for chime warning and locking, and for the chime fly.

Picking up the trail again, Figure 113 shows that the bevel gear set (51) connects with the intermediate chime wheel III (52), a large wheel at the end of a plain arbor. Rotation moves to the intermediate chime wheel IV (53). This wheel is mounted on a short arbor, and is easy to spot. Now we have the connecting point. The link is the chime drive wheel (54), mounted on the same arbor as the chime third wheel (47 in Figure 112).

TRIPLE CHIME SELECTOR

The triple chime selector, chime silencer, and strike silencer are related parts, so we will cover them together. Figure 114 highlights the portion of the upper front movement which concerns us.

The chime selector consists of a lever (42) which pivots, moving the chime shift lever (34) from side to side. There are four positions, corresponding to the three chime melodies and "silent". To hold the lever in each position selected, the chime selector spring (33) slides along the chime selector cam (32). The pin barrel guides, left (55) and right (56) slide the pin barrel laterally on its arbor. This motion brings one of three sets of pins, corresponding to a chime melody, to bear under the hammers.

Figure 114 shows the fourth position, for "silent".

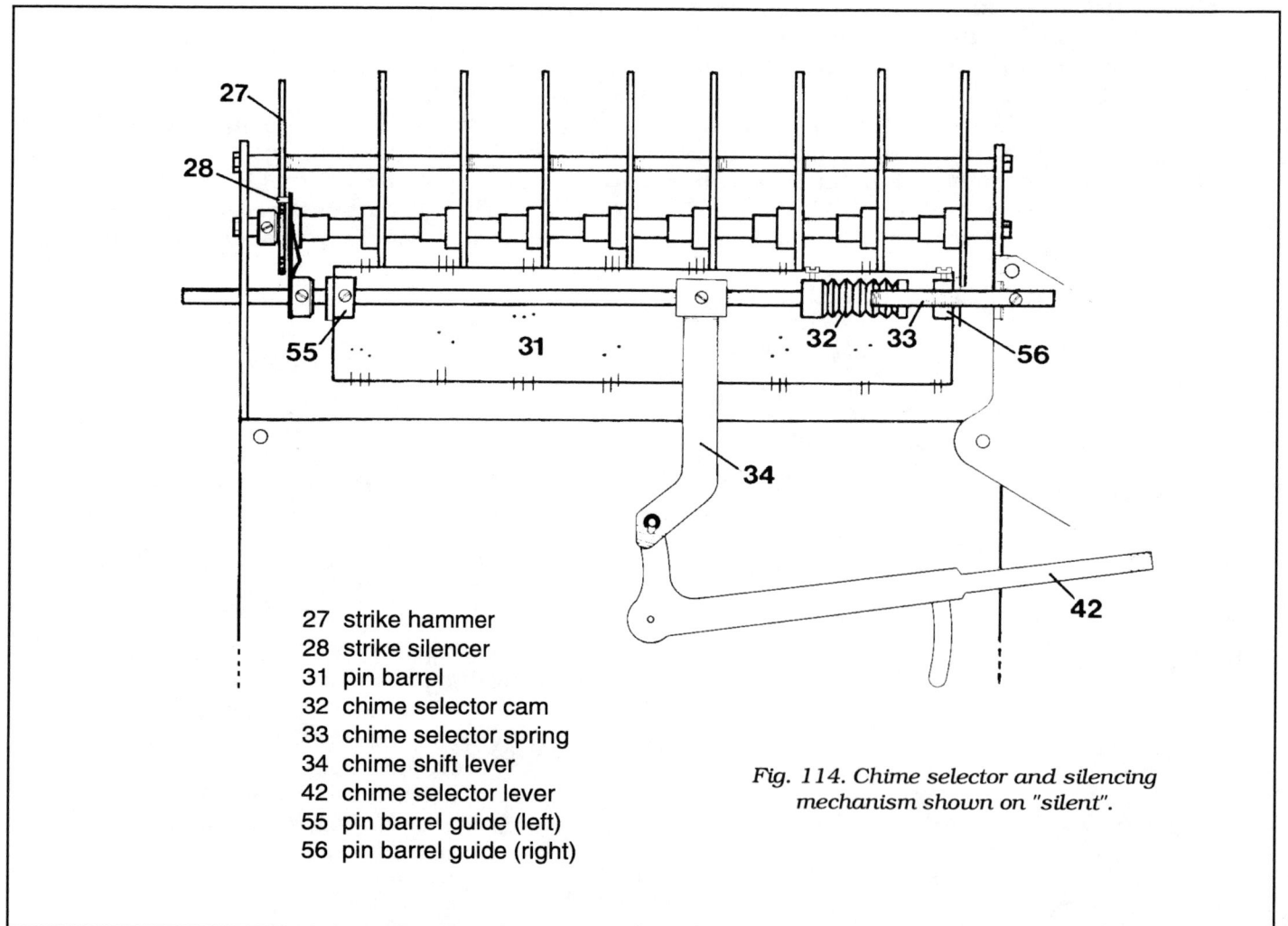

27 strike hammer
28 strike silencer
31 pin barrel
32 chime selector cam
33 chime selector spring
34 chime shift lever
42 chime selector lever
55 pin barrel guide (left)
56 pin barrel guide (right)

Fig. 114. Chime selector and silencing mechanism shown on "silent".

CHAPTER 16 - URGOS 9-TUBULAR BELL

Both chimes and strike are silenced. The chime hammers do not operate, although the chime train still runs each quarter hour. There just aren't any pins on the pin barrel lined up with the hammers, so the notes do not sound.

The strike train also runs on "silent", but the strike hammer (27) is held in the raised position. The strike silencer (28) is pushed to the left as the lever is moved up to "silent" as shown in Figure 114. As the hammer comes back for the first stroke, a spring loaded pin on the silencer catches the hammer and prevents it from hitting the tube. It stays this way until the selector is moved to one of the chime melodies. It is not possible to silence the chimes and strike separately. It is a good feature, however, to have the gear trains run on "silent", for the weights continue to come down together during silent operation.

Check to make sure that the rows of pins are well centered under the hammers as the pin barrel is moved from one chime to the next. To adjust the lateral position of the pin barrel, loosen the set screws on the pin barrel guides (55 and 56). Adjust them so the pins will line up with the hammers, yet remain clear of them during "silent" operation.

A COMMENT ON CHIME SPEED

A common repair problem in the Urgos 9-tube movement is that the chimes are slow. They frequently stall out unless the Westminster, with the fewest notes and smallest hammer load, is selected, and they sometimes stall even then. Various remedies have been tried, but there is no single answer.

There may be several causes. In an eight-note chime sequence, three hammers can be under load at the same time, placing a strain on the chime train. In addition, the chime mechanism has a long path followed by more gears and pinions than are present in other chime clocks. The net result is that the Urgos 9-tube movement, as described in this chapter, may develop the slow-chiming problem after a number of years in service, when a certain amount of wear has occurred in the movement. Experience has shown that complete disassembly and overhaul do not always take care of the problem.

When these problems were encountered in Urgos 9-tube movements in clocks by Colonial Mfg. Co., I was able to restore chiming speed by completely backing off the hammer-spring set screws on the tube rack. The screws were turned out until they had no effect at all on the hammers. The chime volume was still acceptable, and the reduced load on the gear train relieved the problem in the few instances when I tried this remedy.

Other suggestions are to check for worn pivot holes and lack of lubrication. Adding extra weight to the chime side is an option, but added weight rarely cures problems.

This topic brings up the question as to whether some, or all, modern movements have a shorter useful life than those in antique clocks. Arguments can be made about the effects of thin movement plates, poor pivot finish, and possible design flaws in modern movements. I think a combination of factors may result in modern movement problems such as the Urgos chime problem just described.

17
KIENINGER KSU/RSU

The Kieninger KSU/RSU movement is a triple chime, rod chime movement with cable drive. Sized for large grandfather clock cases, this model has the features which are popular today: cable drive with pulleys, triple chimes, fancy lyre pendulum, second hand, and moon dial. There are two similar Kieninger movements which we will consider together. Figure 115 shows the KSU, the higher-cost version. The RSU is the same for repair purposes, but does not have the night shut-off feature, the maintaining power mechanism, or the highly polished plates of the KSU.

This chapter will cover basic movement operation and adjustment and will present hints on assembling the movement. There are also sections on the triple chime shift mechanism and the automatic night chime shut-off feature.

OPERATION OF CHIME AND STRIKE

Let's begin by looking at the front of the Kieninger movement, shown in Figure 116. A locking plate (13) controls the chime train. We will have to look behind the plate to find the cams and levers for chime locking and chime correction. A glance at the strike train reveals a rack and snail arrangement like others we have seen. The strike silencer lever (1) is noteworthy because it works differently than the silencer on most other modern movements. The lever simply rotates the hammer arbor (2), raising the four strike hammers out of the way. At the hour, the strike train runs whether the silencer is activated or not. The chime silencer operates in almost the same manner. When the chime selector lever (14) is moved to the lowest position, the pin barrel is moved forward as far as it will go. The pins will not touch the hammer levers as the chime train runs. The difference, compared to other clocks, is that the chime train continues to run normally when the chimes are silenced: the hammers do not operate.

Figure 117 is a detail of the upper front part of the movement. The plate is cut away to reveal the chime and strike warning wheels and pins. In addition, the long lever across the front of the movement is cut to show that the chime warning lever (11) passes through a rectangular opening in the front plate, enabling it to catch the warning pin. The strike warning lever (18) also protrudes through an opening in the plate. The rack hook and rack have been removed to reveal the gathering pallet (15). Note the locking face on the gathering pallet, which engages a pin on the rack hook to stop the strike train at the end of the cycle.

To understand how the Kieninger movement controls chime locking and chime correction, look at Figure 118. The two views are from the rear of the movement, as though the back plate had been removed. The drawings show only the levers and cams, not the chime train wheels. In Figure 118a, the chime mechanism is in the normal lock position. The chime locking lever (21) is engaged in the slot in the chime locking cam (23), stopping the gear train. Although shaped the same, the chime correction lever (22) has nothing to do with normal locking. It does not engage the slot in the locking cam. The chime correction cam (24) holds the chime correction lever up and out of the way.

Figure 118b shows what happens at the third quarter locking position. The slot in the chime correction cam permits the chime correction lever to move downward, where it is in a position to engage the chime locking cam. This is where automatic chime correction comes into play. Referring briefly to Figure 116, note that the star cam (9) raises the lifting piece (5) each quarter hour. At any quarter, the lift is enough to release the chime locking lever. But when the chime correction lever is activated as in Figure 118b, it takes extra lift for the tab on the locking lever, indicated by an arrow, to be able to raise the correction lever high enough to clear the locking cam. Only one arm of the star cam, for the hour, is long enough to provide the high lift.

CHAPTER 17 - KIENINGER KSU/RSU

Fig. 115. Kieninger Model KSU, triple chime movement with night shut-off feature. Photo Kieninger and Carl Baessler, manufacturer's representative.

Moving to the back of the Kieninger movement, we can look at the rest of the chime and strike parts in Figure 119. On the strike side, the plate is cut out to reveal the hammer-lifting star (32), which raises the strike hammer lever (34) and the four strike hammers (31). The hammers are shown in the rest position, with the hammer tail and hammer-lifting star free of each other. As with any strike train, the hammers should not be expected to start up under the load of the hammers. The adjustment is described in the next section. The chime parts in Figure 119 are shown with the bracket plate removed. The chime drive wheel (27) turns the pin barrel (26). The pins lift the eight hammer levers (25), and the hammers (30).

ASSEMBLING THE MOVEMENT

The winding drums and main wheels should be left aside as you fit the plates together. As with many other chime movements described in this book, begin by inserting all the arbors into the front plate. Now add the rear plate and fit all the pivots into their holes. Install the gathering pallet, rack hook, chime levers, lifting piece, and locking plate. It is difficult to assemble the gear trains with all the wheels, pins, and levers in their correct positions, so you should expect to separate the plates slightly, to adjust the relationship of the parts.

For the chime, the first adjustment is to set the warning pin correctly. With the chime locking lever engaged as in Figure 118a, make sure you have nearly a half turn of warning run, shown in Figure 117. It may take several tries to lift the back plate and correctly mesh the chime warning wheel pinion and the locking wheel below it. Now turn the wheels until the chime correction lever drops into the slot in the chime correction cam. If the chime locking lever drops to engage the locking cam at this point, the third quarter locking point is established. Loosen the locking plate and move it to the third quarter locking point, shown in Figure 116. If the locking cam slot does not present itself at the right moment, change the mesh of the pinion on the locking arbor with the wheel below.

Make your strike adjustments next. If the train locks with the warning pin (17) in position as indicated in Figure 117, you will have the correct strike

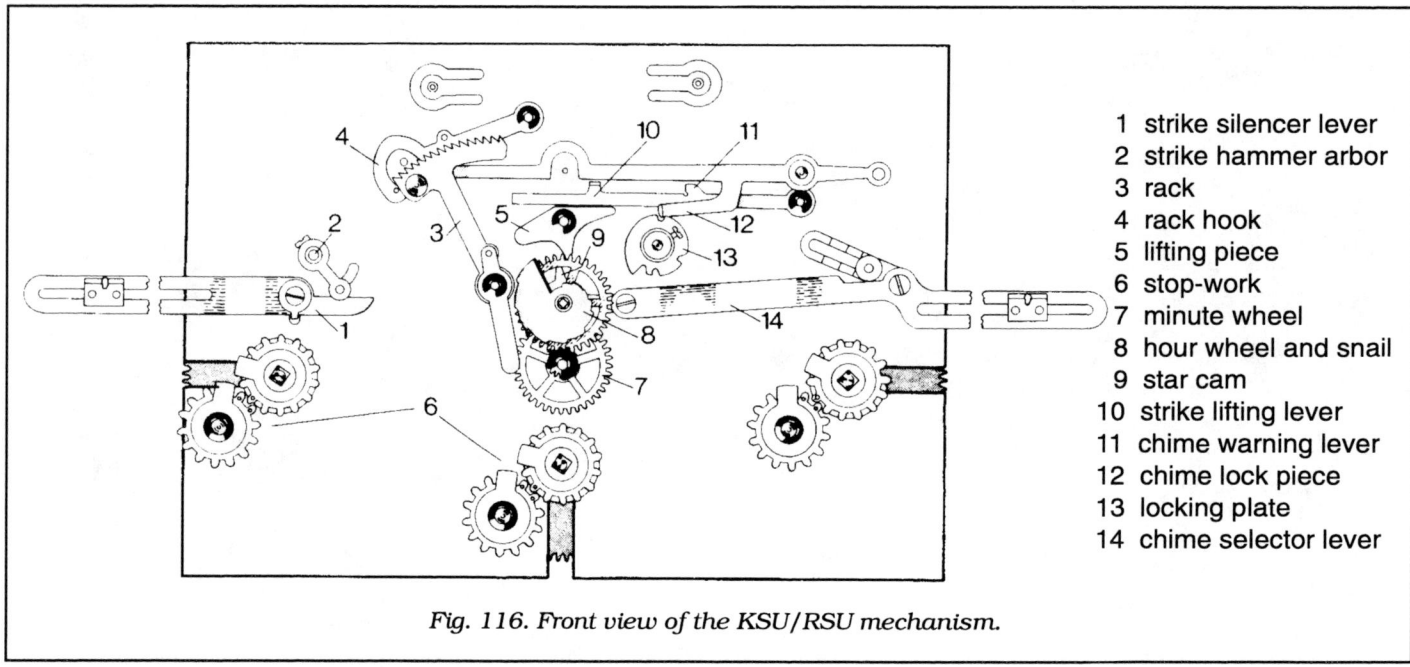

Fig. 116. Front view of the KSU/RSU mechanism.

1 strike silencer lever
2 strike hammer arbor
3 rack
4 rack hook
5 lifting piece
6 stop-work
7 minute wheel
8 hour wheel and snail
9 star cam
10 strike lifting lever
11 chime warning lever
12 chime lock piece
13 locking plate
14 chime selector lever

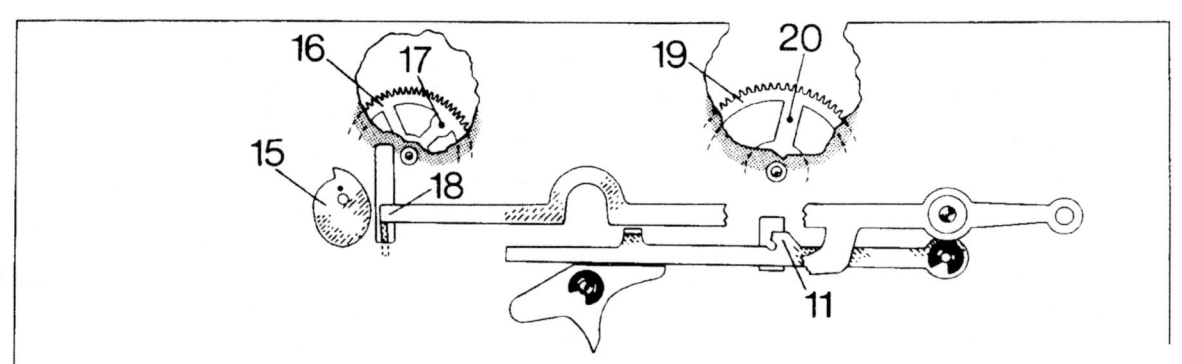

Fig. 117. Detail of front movement chime and strike parts, with rack, snail, and locking plate removed.

11 chime warning lever
15 gathering pallet
16 strike warning wheel
17 strike warning pin
18 strike warning lever
19 chime warning wheel
20 chime warning pin

CHAPTER 17 - KIENINGER KSU/RSU

warning run. Change the relationship of the warning pinion with the gathering wheel below to correct the adjustment. Now check to be sure the hammer is free of the hammer-lifting star when the strike train locks. You will probably have to make an adjustment. Ease the arbor carrying the hammer-lifting star out of its hole, and move it to a new position. Check the hammer operation again. Add the rack and snail, making sure that you check them at 12 and 1 o'clock.

After the movement is assembled, slide each winding drum into its slots and then attach the front and rear pivot plates. The movement is fairly easy to handle as a result of this feature. The stop-work (6), shown in Figure 116, can be added at the very end, when the movement is running on the test stand. Wind the pulleys up as high as they should go, near the seatboard. Then install the gear to the left of each winding arbor. Line up the punch marks stamped on each gear, as indicated on the drawing, and the stop-work will be properly adjusted. Its function is to prevent over-winding.

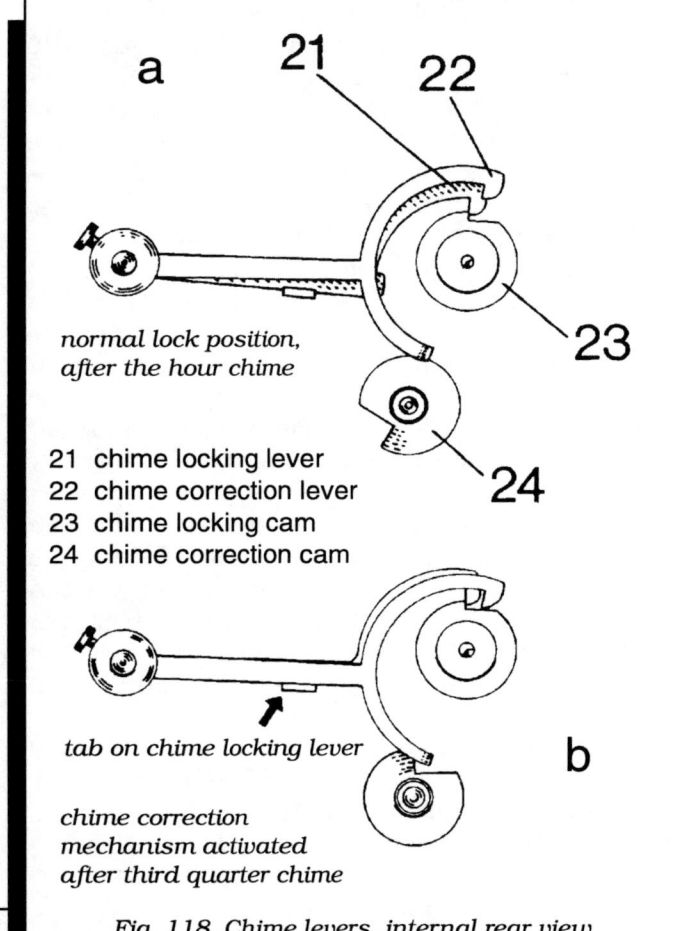

a

normal lock position, after the hour chime

21 chime locking lever
22 chime correction lever
23 chime locking cam
24 chime correction cam

tab on chime locking lever

b

chime correction mechanism activated after third quarter chime

Fig. 118. Chime levers, internal rear view.

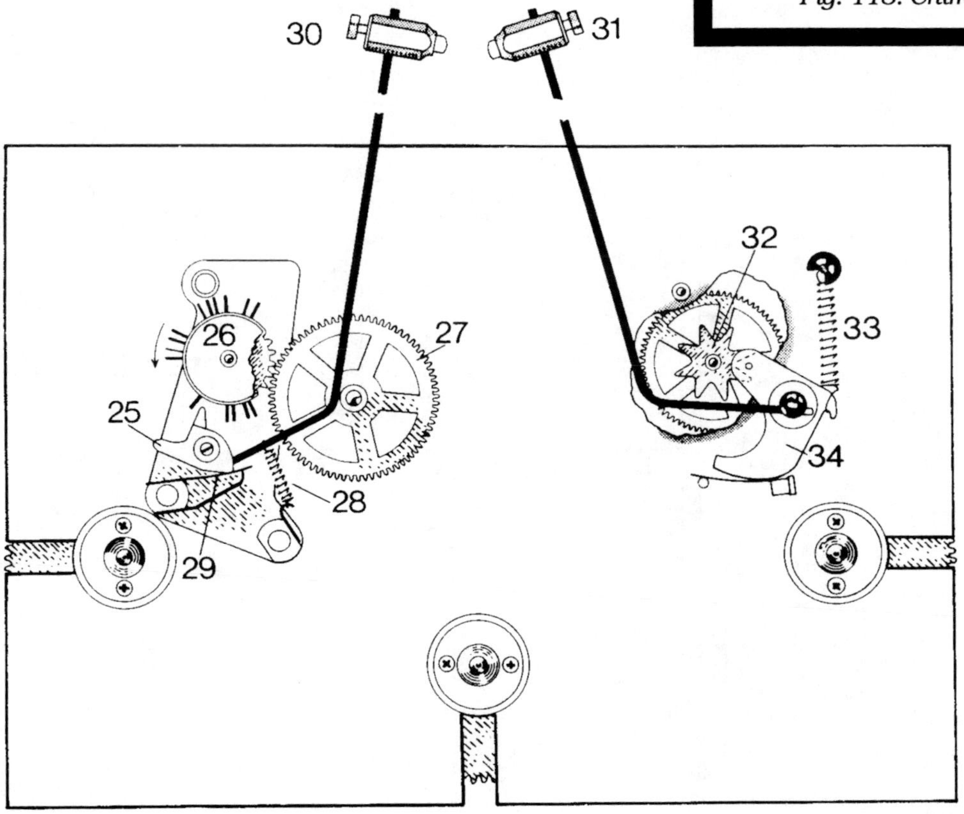

Fig. 119. Rear movement chime and strike parts, chime bracket plate removed.

25 chime hammer levers (8)
26 pin barrel
27 chime drive wheel
28 chime hammer springs (8)
29 buffer springs (8)
30 chime hammers (8)
31 strike hammers (4)
32 hammer-lifting star
33 strike hammer spring
34 strike hammer lever

TRIPLE CHIME SHIFT MECHANISM

The triple chime mechanism shown in Figure 120 must be carefully adjusted if the Kieninger movement is to play each chime and also be capable of silent operation. In this mechanism, the precise front-to-back position of the pin barrel (26) is controlled by the pin barrel lever (49). As the chime selector lever (14) is moved up or down, the cam (41) pushes the chime selector arbor (48) correspondingly forward or back. The pin barrel lever is set-screwed to the arbor, and moves with it. The lever fits in a recess in the pin barrel, moving it along to each position. Pressure from the chime selector spring (47) keeps the end of the chime selector arbor pressed against the cam, assuring positive action.

If the set screw on the pin barrel lever has been loosened during repairs, the lever will have to be adjusted again. Try different settings until the pin barrel slides to the correct position for each chime. When the row of pins for one of the chimes lines up squarely under the hammer tails, the others will also be correct.

The bottom position of the chime selector lever is for "silent". This places the pin barrel as far forward as it can travel. Observe that none of the pins will contact hammers at this setting. The chime train still runs when on "silent", but the pin barrel doesn't lift any hammers. Most chime trains remain stopped during silent operation, but the Kieninger runs. A benefit of this arrangement is that the weights continue to descend evenly during silent periods. Remember, however, that putting the clock on silent does not make it safe for you to turn the minute hand rapidly clockwise. The chime train will still be running.

Moving the chime selector lever upward will bring in Westminster, St. Michael, and Whittington chimes. As you select each one, double check that the pins touch the hammer tails instead of falling on the edges or in between them. Also, adjust the pin barrel so the correct notes are played at each quarter. Turn the chime train through the first quarter chime, then loosen the set screw on the chime drive wheel. Move the wheel and pin barrel together, independently from the chime train. When you observe the eight hammers (on one of the eight-note chime melodies) operating from rear to front, tighten the set screw again.

There are a few areas to lubricate and test on the triple chime shift mechanism. Make sure the lever can be moved up and down smoothly, without forcing. Use a light grease on the cam to assure smooth operation. Check the spring to be certain

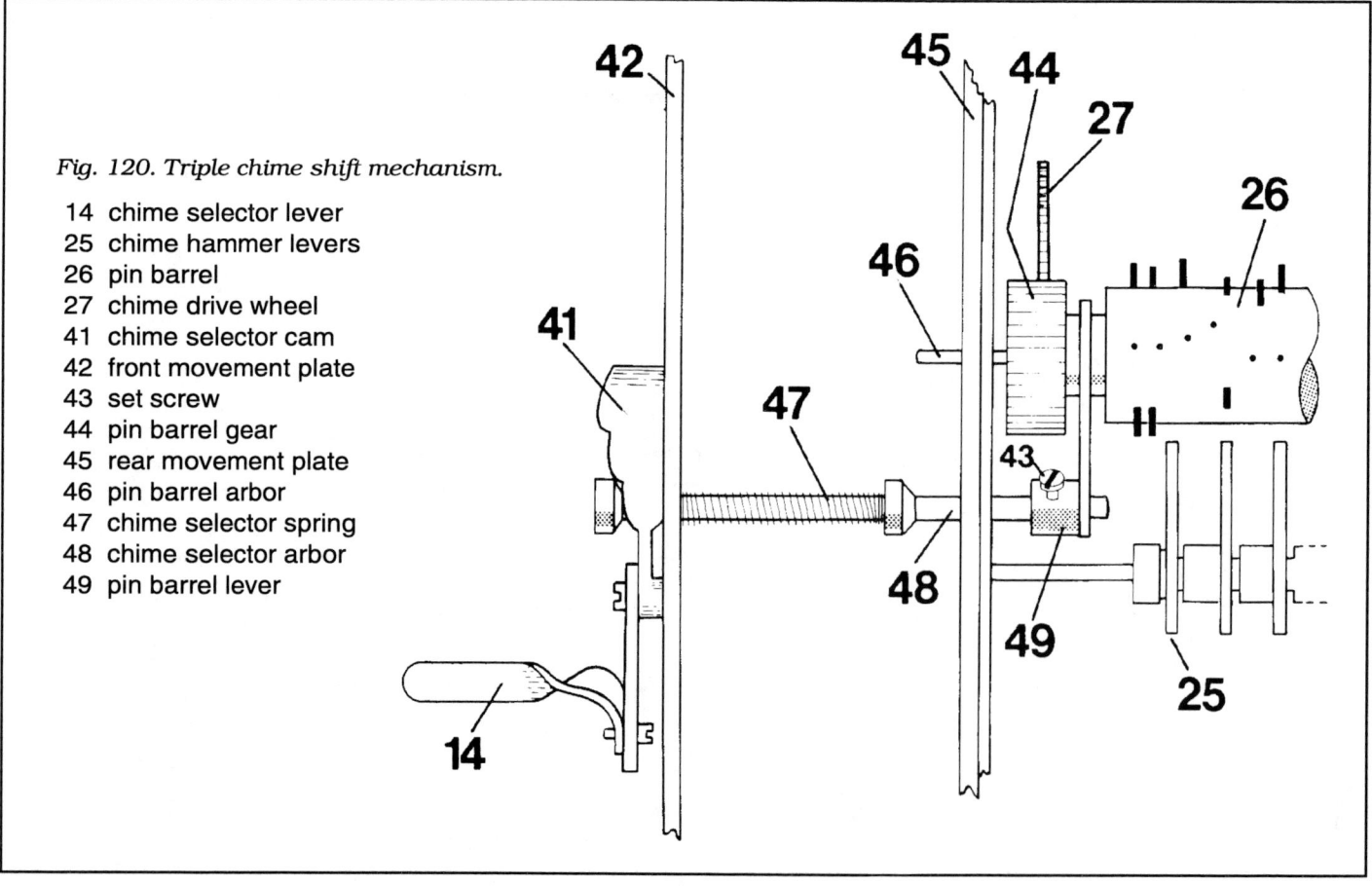

Fig. 120. Triple chime shift mechanism.

- 14 chime selector lever
- 25 chime hammer levers
- 26 pin barrel
- 27 chime drive wheel
- 41 chime selector cam
- 42 front movement plate
- 43 set screw
- 44 pin barrel gear
- 45 rear movement plate
- 46 pin barrel arbor
- 47 chime selector spring
- 48 chime selector arbor
- 49 pin barrel lever

CHAPTER 17 - KIENINGER KSU/RSU

that it exerts positive pressure on the mechanism. After you install the dial, look at the chime selector lever from the front. It should not press against either side of the slot in the dial. If it does, the lever will be hard to move. In addition, you cannot feel the lever click into place for each chime position if the lever rubs on the slot. Adjust the placement of the lever to center it in the slot. Finish up by oiling the pivot holes for the pin barrel arbor and the chime selector arbor, and add some light grease to the contact surface between the pin barrel lever and the recessed portion of the pin barrel.

AUTOMATIC NIGHT SHUT-OFF

Refer to Figure 116, which shows the strike silencer lever as a direct means of lifting the strike hammers out of the way to silence them. The strike train still runs in the "silent" position, but the hammers are raised away from the rods. Moving now to Figure 121, the night shut-off selector lever (35) replaces the simpler lever. The new lever still permits you to silence the strike in the same way as before. It also operates the night shut-off mechanism shown in Figures 121 through 124.

Perhaps it is best to begin by explaining what the night shut-off feature does. When you select it by moving the lever, the mechanism automatically silences chime and strike each night. Kieninger literature states that "standard settings" are either 8 or 9 full hours of silencing, depending on how the unit is made. The KSU movement I used to prepare this chapter stops chiming and striking for eight hours, following the music at 11:00. The clock automatically resumes chiming and striking at the end of the period. A clock owner who works at night and sleeps by day could easily make the silent period fall during the day instead, by turning the clock ahead exactly 12 hours.

Power for the mechanism comes through the hour wheel and snail (8) and the minute wheel (7). The drive wheel (40) transmits motion to the night shut-off cam (36). Because of the geared relationship between the snail and this cam, the beginning of the silent period is fixed. You can change it only by removing the cam and putting it back on at a different orientation. Duration of the silent period is a different matter. It depends entirely on the length of the raised arc on the cam.

The night shut-off selector lever has three positions. These are labelled "night off" (middle), "strike" (top), and "silent" (bottom). We will cover each position in detail, but it is a good idea to summarize them first. The middle position places the night shut-

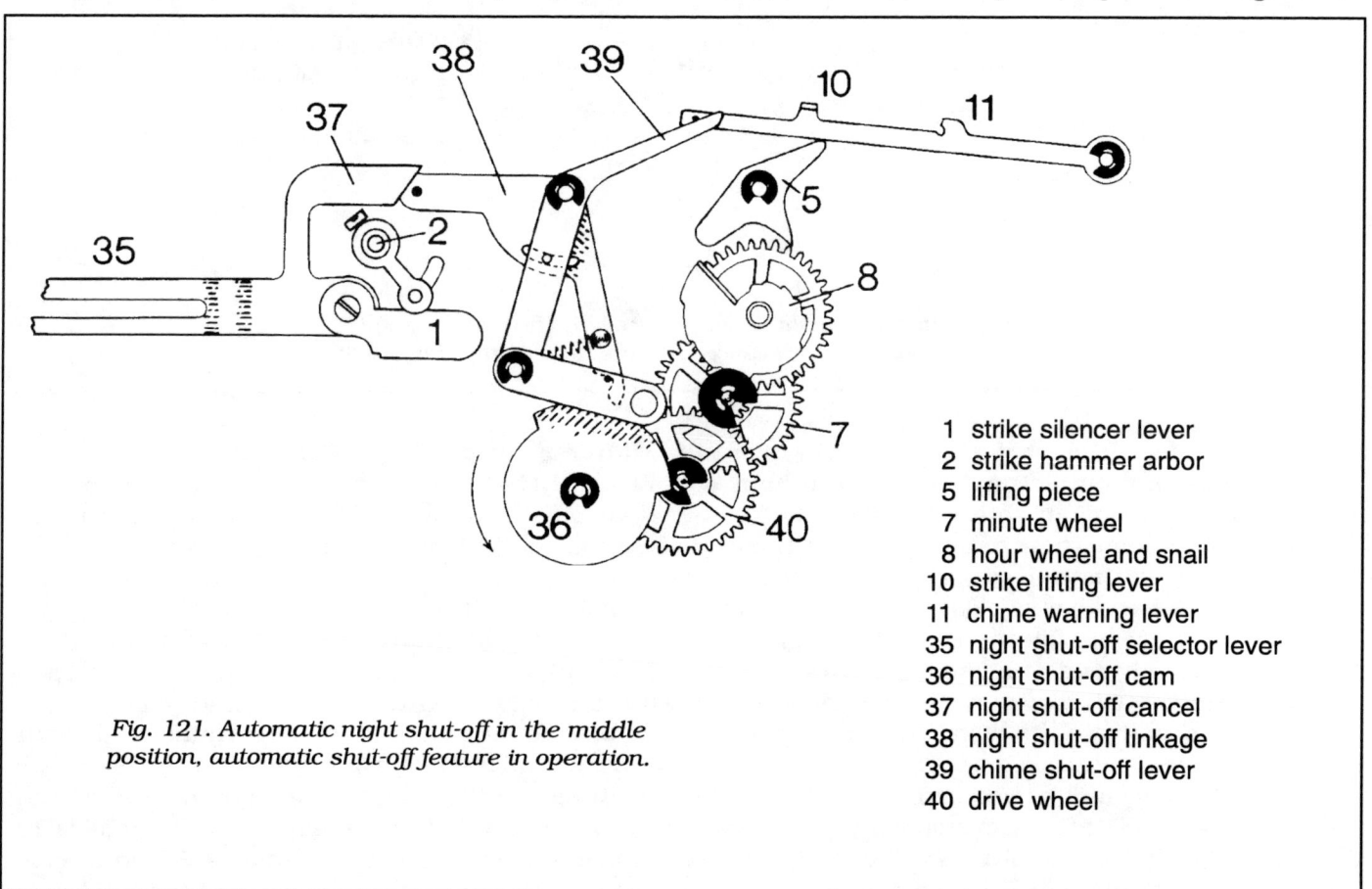

Fig. 121. Automatic night shut-off in the middle position, automatic shut-off feature in operation.

1 strike silencer lever
2 strike hammer arbor
5 lifting piece
7 minute wheel
8 hour wheel and snail
10 strike lifting lever
11 chime warning lever
35 night shut-off selector lever
36 night shut-off cam
37 night shut-off cancel
38 night shut-off linkage
39 chime shut-off lever
40 drive wheel

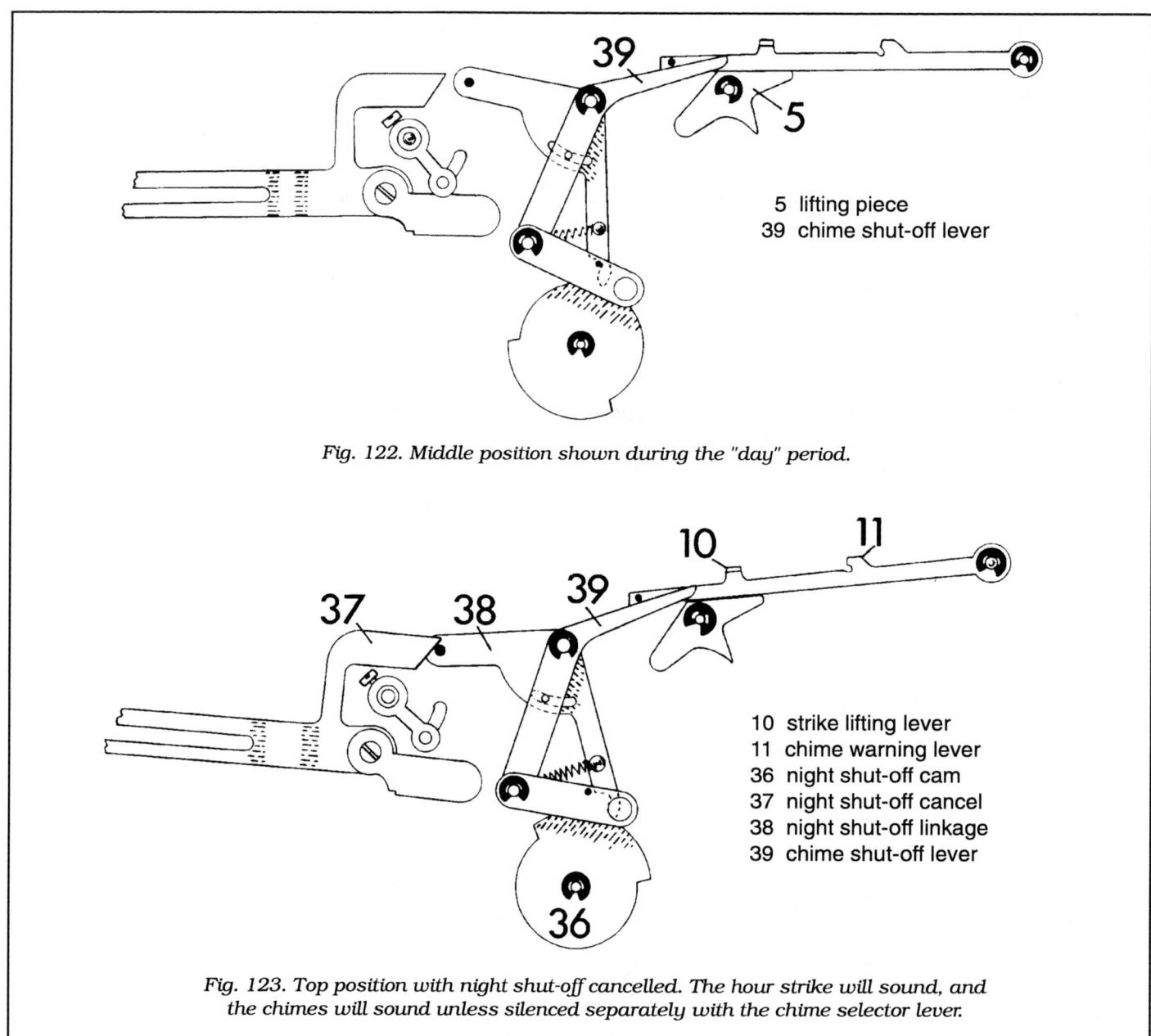

Fig. 122. Middle position shown during the "day" period.

Fig. 123. Top position with night shut-off cancelled. The hour strike will sound, and the chimes will sound unless silenced separately with the chime selector lever.

off feature in effect. The top position deactivates or cancels it. The bottom setting silences the hour strike but leaves the night shut-off in effect.

When you combine these options with the three chimes plus "silent" that you can select on the chime selector lever on the other side of the dial, you have great flexibility. It is possible to silence either chimes or strike separately and still choose whether you want to hear the music play at night. One thing you cannot do is deactivate the night shut-off and silence the strike at the same time. In other words, if you want 24-hour chiming you cannot silence the strike. Explaining all this to a customer might prove difficult. Fortunately, most owners will settle on the settings they like after a little experimentation.

Middle Position "Night Off"

Figure 121 shows the middle position of the night shut-off selector lever (35). The automatic shut-off is in effect. The night shut-off cam (36) raises the chime shut-off lever (39). In turn, the lever lifts the chime lever (10 & 11). This sends the chime train to the warning position, where it remains throughout the shut-off period. The strike train does not operate. During the night, the chime and strike weights do not move down, and the gear trains do not run.

After the last hour of silencing has passed, the night shut-off cam has turned far enough to bring the other part of the cam beneath the shut off lever. The chime lever (10 & 11) is now free to move. At the next quarter hour, the chime train is released

CHAPTER 17 - KIENINGER KSU/RSU

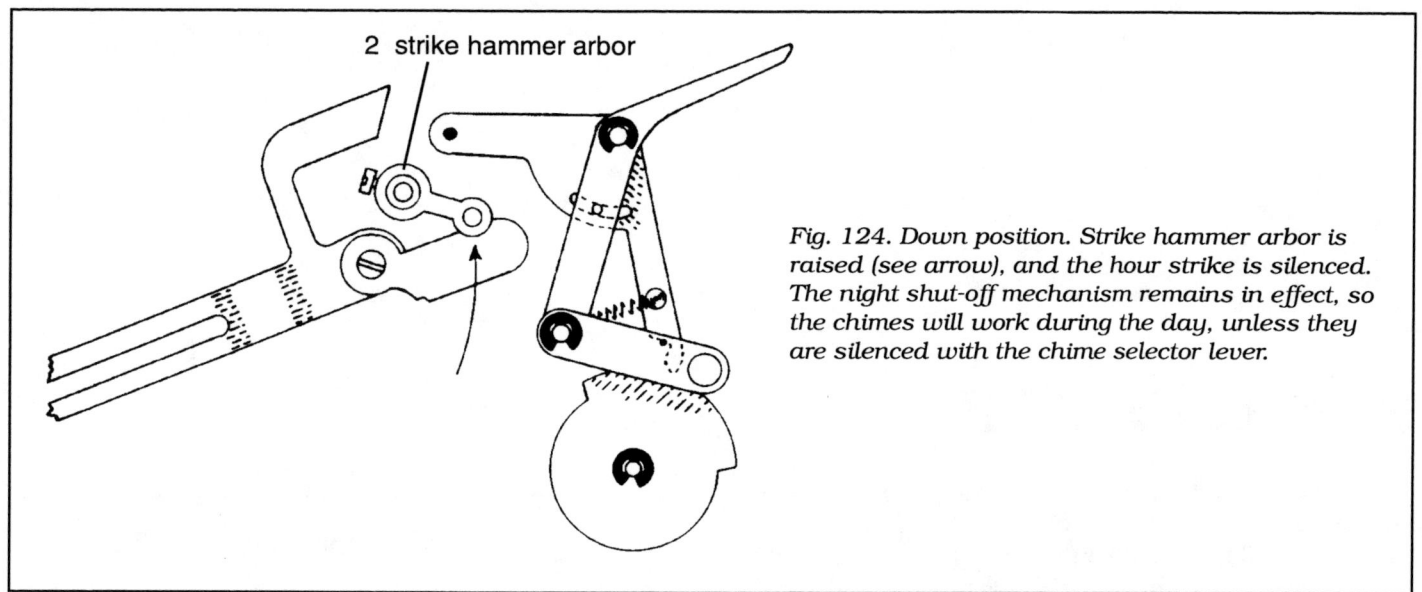

Fig. 124. Down position. Strike hammer arbor is raised (see arrow), and the hour strike is silenced. The night shut-off mechanism remains in effect, so the chimes will work during the day, unless they are silenced with the chime selector lever.

from warning, and chiming resumes. Figure 122 shows the mechanism at about the midpoint of the "day" period. The lifting piece (5) does its job each quarter, and there is no interference from the chime shut-off lever (39).

Top Position "Strike"
Refer to Figure 123 to see what happens when you move the night shut-off selector lever up. The night shut-off cancel (37) pushes on the night shut-off linkage (38). The linkage pivots against spring pressure. The hooked bottom part of the linkage, represented by dotted lines just above the cam (36) in the figure, moves out of the way. As a result, the nearby pin on the chime shut-off lever, shown in black, cannot exert force. It's as though the mechanism had been partly disassembled: the parts still move, yet nothing happens.

Figure 123 shows the cam in the night position, but you can see that the shut-off lever (39) has not affected the chime lever (10 & 11). The chime train will operate each quarter hour, day and night. Even if the chime selector lever on the other side of the movement has been placed on "silent", the chime train will still run. The hammers would not operate, however. Under this setting, the strike will be heard each hour. This means that even though the chimes may be silenced, the strike will sound. Both chime and strike gear trains operate. The clock owner can select one of three chime melodies, or silence the chimes, but he will hear the strike each hour.

Bottom Position "Silent"
All the selector does in this position is to mechanically raise the strike hammers to silence them. Figure 124 shows the strike silencer lever pushing up on the strike hammer arbor (2). The night shut-off feature remains operable. The hammers are raised, so there is no strike despite the fact that the strike train runs during the "day" period. And during each "day" period, the chimes will sound unless they have been silenced separately. During the night, the mechanism stops the chime and strike trains from running.

To understand the night shut-off feature, you should first understand the strike silencer is just a device to raise the strike hammers. It has no effect on the gears, because the strike train continues to run each hour whether it is silenced or not. Encourage Kieninger owners to experiment with the night shut-off and triple chime features until they find one or more settings which meet their needs. Most people don't continue to change chime melodies and silencing combinations very often, once they have found what they like best.

18

GEBR. JAUCH

The Jauch movement shown in Figure 125 was used by several U.S. clock companies for wall and shelf clocks. Colonial Manufacturing used four versions of the movement with various pendulum lengths for shelf clocks as late as 1982. The example studied for this chapter is a No. 41 2800/B, stamped with the number "68" on the back plate.

Before going through the repair procedure, we will take a quick survey of the movement's features. It is Westminster chime, with five hammers mounted underneath the movement. The strike is rack and snail type. The chime train is locking plate controlled. There is a chime correction cam on the same arbor as the locking plate. It is located between the movement plates rather than on the back of the locking plate as we are more used to seeing it in chime clocks. The chime lock pin serves a dual role in this movement. It extends both to the front and rear of the locking wheel. The front portion is actually the chime lock pin, and the rear extension is the chime correction pin. Strike and chime trains have a fixed warning run from the locking lever to the warning lever, with the same pin handling locking and warning. This means the warning pin does not have to be set as in many other movements.

DISASSEMBLY

Following an examination of the movement for any unusual wear or damage, you are ready to take it apart. First, let down the three mainsprings. The clicks are mounted on the rear of the winding arbors, inside the rear plate. Refer to Figures 126 through 128 to identify parts as you work with them.

On the rear of the movement, there are several parts to be removed. The parts are arranged in a different manner than other movements, so a diagram (Figure 128) is needed for reference. Remove the bracket plate (26) and the strike linkage (23). Take out the two idler wheels (25). Unscrew the hammer shaft (28) and take off the hammers and shaft as a unit. Note that a spacer bushing is lo-

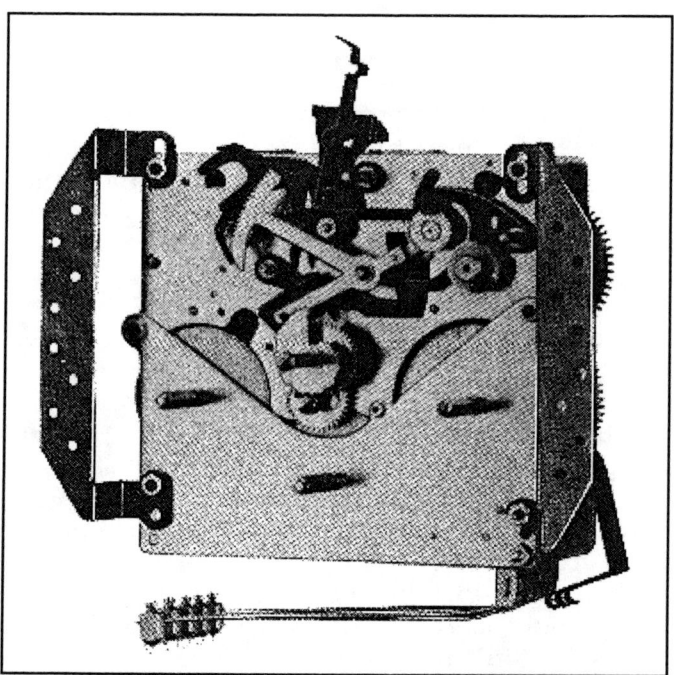

Fig. 125. Gebr. Jauch chime movement No. 41-2800/B.

cated on the back end of the hammer shaft. Remove the chime drive wheel (22) after backing off the two set screws. The wheel may be very tight on the arbor if burrs have been raised by over-tightening of the set screws. Carefully pry off the wheel using two levers to apply equal force on opposite sides. Polish off the burrs in the lathe later on. Even before removing the arbor and wheel from the back plate, you may have to file off a burr which prevents the pivot from going through the hole. Note that there is a long spacer bushing for the lower pillar extension, under the bracket plate.

The front movement parts can be removed next. Referring to Figure 126, take off the locking plate (5), the black-finished levers (2, 3, and 6) behind it, and the rack hook (14). Remove the movement brackets to get them out of your way, and put the upper pillar nuts back on finger tight. Take off the

CHAPTER 18 - GEBR. JAUCH

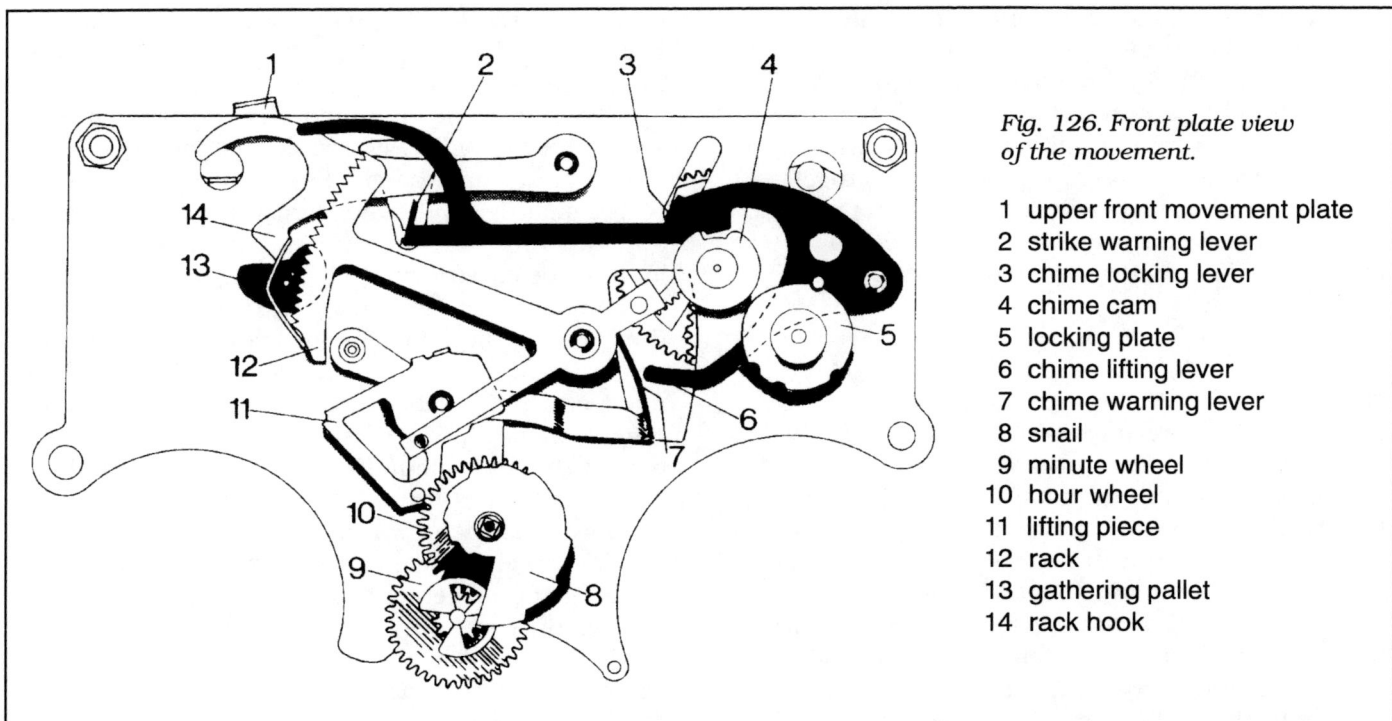

Fig. 126. Front plate view of the movement.

1 upper front movement plate
2 strike warning lever
3 chime locking lever
4 chime cam
5 locking plate
6 chime lifting lever
7 chime warning lever
8 snail
9 minute wheel
10 hour wheel
11 lifting piece
12 rack
13 gathering pallet
14 rack hook

lower front movement plate and the barrels. Mark the barrels for later identification. The click wheels are all the same. They may be difficult to remove from the squared ends of the winding arbors. Before removing the upper front movement plate, mark the time second wheel and chime fly for identification. The chime correction lever (18) and the chime warning lever (7) are held in place by split washers between the plates: these may be left in place during ultrasonic cleaning. Drive off the chime cam (4) after the plates are separated. This is the safest way, better in this case than using levers. Twisting will break the pivot. Proceed to remove all parts and proceed with cleaning.

REASSEMBLY AND ADJUSTMENT

Bushing work, pivot polishing, and other repairs will be necessary. Reassembly now follows. Assemble the arbors into the front plate, leaving the barrels and lower front movement plate aside for the mo-

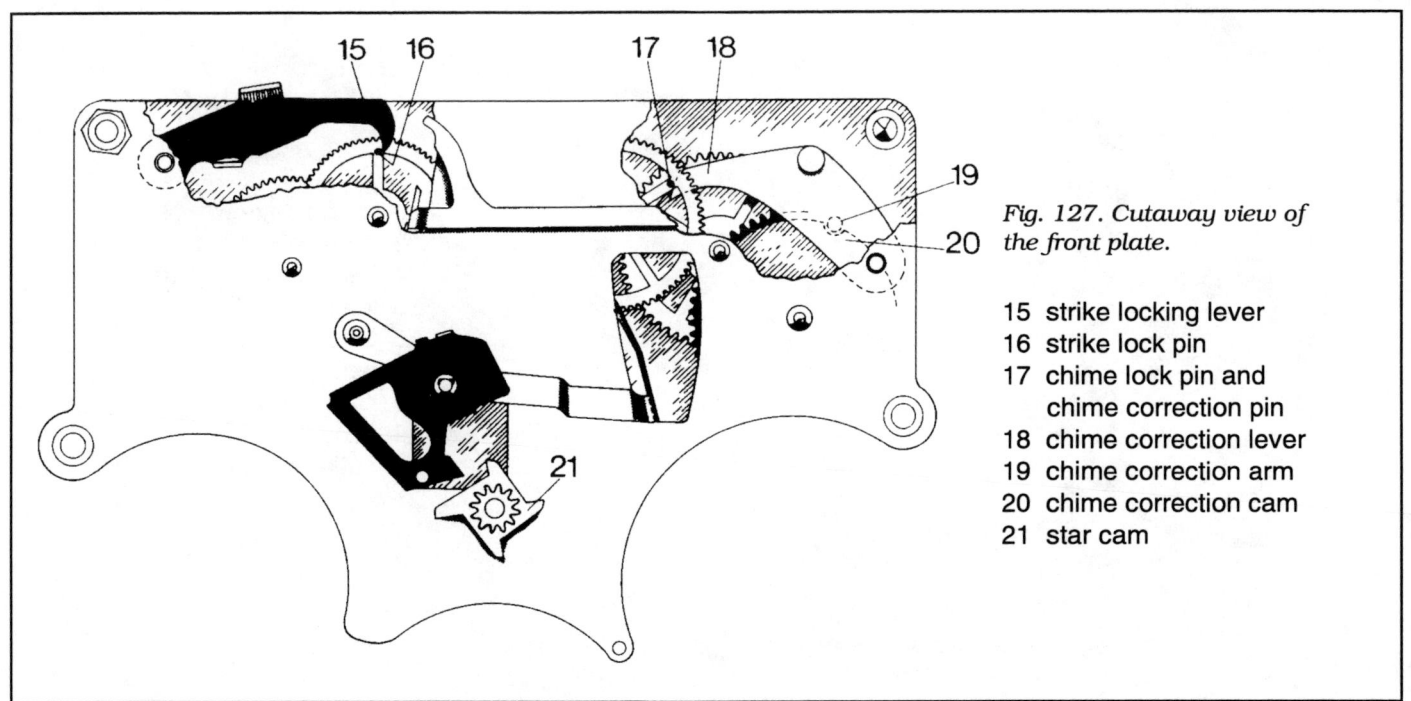

Fig. 127. Cutaway view of the front plate.

15 strike locking lever
16 strike lock pin
17 chime lock pin and chime correction pin
18 chime correction lever
19 chime correction arm
20 chime correction cam
21 star cam

ment. The suspension unit and pallets should also be left aside. Do not forget to add the strike locking lever (15), which goes between the plates. Watch out for the chime correction arm (19) because it may jam against the chime correction cam (20) as you try to bring the front and back plates together. This will hold the plates apart, causing you to search about for a jammed pivot. Never force clock plates together: always locate and correct whatever is keeping them apart.

Strike

After the plates are together, do not proceed with all the rest of the front and rear movement parts until strike and chime adjustments have been made. The first step is to work on the strike train. Add the long lever which includes the strike warning lever (2), and also add the rack hook (14). The strike train can now be moved by finger pressure, if you release the rack hook. The fly rotates counterclockwise as viewed from the front. Each time you release the rack hook, the gathering wheel makes one revolution before locking occurs again. Locking should happen as the rack hook reaches the notch in the gathering pallet (13). The strike locking lever (15) stops the strike lock pin (16) at this time.

If the locking and warning wheels are not synchronized, the locking will take place as the rack hook slides down the slope of the gathering pallet. This causes a shallow lock or even a mis-locking. Note that it is the rack hook, riding the curves of the gathering pallet, which moves the strike locking lever up and down. Separate the movement plates slightly, and synchronize the locking wheel and warning pinion as necessary. The pivots are long, so be careful not to bend them.

The next step is to check the strike hammers. If they are left under load as the train locks, this must be corrected. As with any movement, the hammers must end up at rest. To check the hammers, you must first assemble the strike linkage (23) onto the rear of the movement. The arbor goes through the bracket plate (26) which can be held on with one nut for adjustment purposes. The other pivot point for the strike linkage is another arbor which runs between the movement plates. It is held by a split washer on the front plate, but you can leave the washer off for now. Since the rack is not in place yet, the train will strike only once each time the rack hook is released and the wheels are turned by hand. This is the action to observe. If the hammer is free of the hammer-lifting star (24) then the set-up is correct. If an adjustment is needed, separate the plates and synchronize the pin wheel with the pinion on the gathering (locking) arbor.

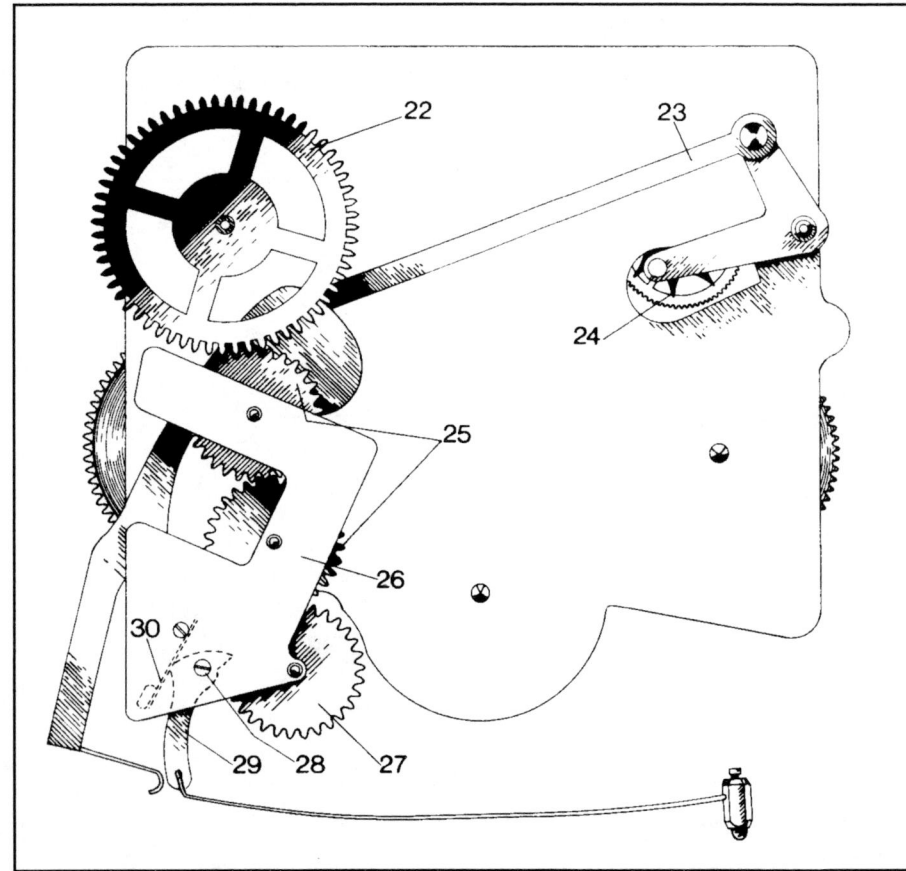

Fig. 128. Back plate.

22 chime drive wheel
23 strike linkage
24 hammer-lifting star
25 idler wheels
26 bracket plate
27 pin barrel gear
28 hammer shaft
29 hammers (5)
30 buffer springs

CHAPTER 18 - GEBR. JAUCH

Chime

The chime cam (4) must be positioned so that when the chime drop lever goes down into the slot in the chime cam, the chime lock pin (17) will be ready to stop the chime locking lever (3). The function of the chime cam, with its single slot, is to hold the chime locking lever away from the lock pin except at those times when locking is supposed to occur. The most critical moment is just as chiming begins.

Adjusting the chime cam is not easy. It is driven tightly onto the arbor, so it is unwise to try to twist the cam to a new position. It is better for you to have driven the cam off during disassembly, as suggested earlier. Now it can be re-installed easily, and not driven on so tightly. As an alternative, it is possible to separate the plates enough to ease the front pivot of the locking wheel out of the hole, but be careful, because the escape wheel is partly In your way. Recheck by rotating the chime train by hand. At each revolution of the chime cam, the train should lock. Next, install the locking plate (5). When the chime train is locked, the chime cam and locking plate work together to permit locking of the mechanism.

Chime Correction Mechanism

The chime correction cam (20) is between the plates, on the same arbor as the locking plate. After the third quarter chime, the chime correction arm (19) drops into the single slot in the chime correction cam, lowering the chime correction lever (18) so that it stops the chime correction pin (17). Only at the third quarter does the locking take place this way. At the other quarters, the chime lock pin and chime locking lever do the job. The slightly longer hour arm on the star cam (21) is the only one of the four arms long enough to lift the chime levers to start the chimes for the hour.

It stands to reason that it is the slot in the chime correction cam which locates the hour and synchronizes the locking plate. Make sure that the chime correction arm (19) drops into this slot just as the third quarter chime sequence ends. If it does not, there will be no chime correction. Since the chime correction cam is held by a tension washer, it can be rotated on its arbor for adjustment. The parts are not synchronized by separating the plates and re-indexing a wheel and pinion. You will not have to move the chime correction cam very far. As long as the slot rotates to a position exactly at the end of any sequence of chime notes, the adjustment will be correct.

FINISHING ASSEMBLY AND ADJUSTMENTS

The precise positions of the chime cam, locking plate, and chime correction cam are critical. To make it easier to finish the adjustments, install the mainspring barrels with click wheels installed on the arbors, and the lower front movement plate. Tighten eight hex nuts on the front of the movement, and make sure everything is secure on the back plate. Wind the strike and chime springs partially. Add the hour wheel (10), the minute wheel (9), and the rack (12). Adjust the snail position for correct striking at 12 and 1 o'clock. Watch the operation of the chime correction cam. If the slot rotates under the chime correction arm too soon, the chime train will stop prematurely, with perhaps one note left to play. If the slot arrives on position too late, there is no chime correction at all. Careful observation and adjustment are required.

After you are satisfied with chime and strike operation, proceed to finish assembling the movement. Install all remaining split washers which hold levers in place. On the rear of the movement, we left the bracket plate held on with one nut. Remove the plate now, and screw in the hammer shaft with the hammers on it. Put in the pin barrel and gear (27), and the idler wheels (25). Add the bracket plate and fasten with the nuts. Adjust the buffer springs (30) to allow the hammers to fall by gravity. The springs cushion the hammer action to prevent excessive bouncing. Add the chime drive wheel (22) and adjust its position for correct chime sequence. On the movement, the hammers should work in order from front to rear for the first quarter chime.

To finish up, let down the mainsprings if you have not installed the movement brackets, and add them now. You recall that we left the brackets out because they get in the way during the assembly of the movement. Install the pallets and suspension unit, wind the mainsprings, and test the movement.

19

MAUTHE W500

During the 1960's and early 70's, spring driven movements were a popular choice in grandfather clocks. The movements fitted well with the small cases being produced to complement the furnishings of the time. Many of these clocks were about six feet in height, and were of the narrow-waisted style.

Figure 129 shows the Daneker "York County", typical of this style of small floor clock. The case is 71 inches tall including the finial, and only 11 inches wide at the waist section. This is too small a clock to accommodate weights or a large pendulum disk. The movement by F. Mauthe of Germany is well matched to the case. It has a small cylindrical pendulum and is spring driven. Hammers are mounted on the back of the movement, with four strike hammers on one side and four chime on the other.

Just as Daneker is remembered for its fine cabinet finishes, many repairers associate the Mauthe movements with broken mainsprings, a frequent cause of movement failure. As late as 1978, Daneker still sold replacement Mauthe barrels complete with mainsprings for the W500. Now the repairer must find the closest replacement springs from suppliers, and damaged barrels must be re-toothed at considerable cost. Customers would often continue to run the Daneker clocks after the hole-end snapped out of a mainspring. They would wind the spring two or three turns, until it began to slip around in the barrel. Winding the clock every two days kept it going, and they didn't call for service until they were tired of winding the clock!

Figure 132 shows a short pendulum mantel and wall version of the Mauthe W500 that was installed in a variety of cases. It is the same basic movement as the floor clock model, but with the hammers mounted in a row underneath the movement as shown in Figures 132 and 133. Repair procedures are similar to the floor clock, but there are some differences which will be covered in this chapter.

Figures 130 and 131 identify the major chime

Fig. 129. Daneker "York County", 71 inches tall, with Mauthe W500 movement. Fine cabinet work and glossy maple finish.

CHAPTER 19 - MAUTHE W500

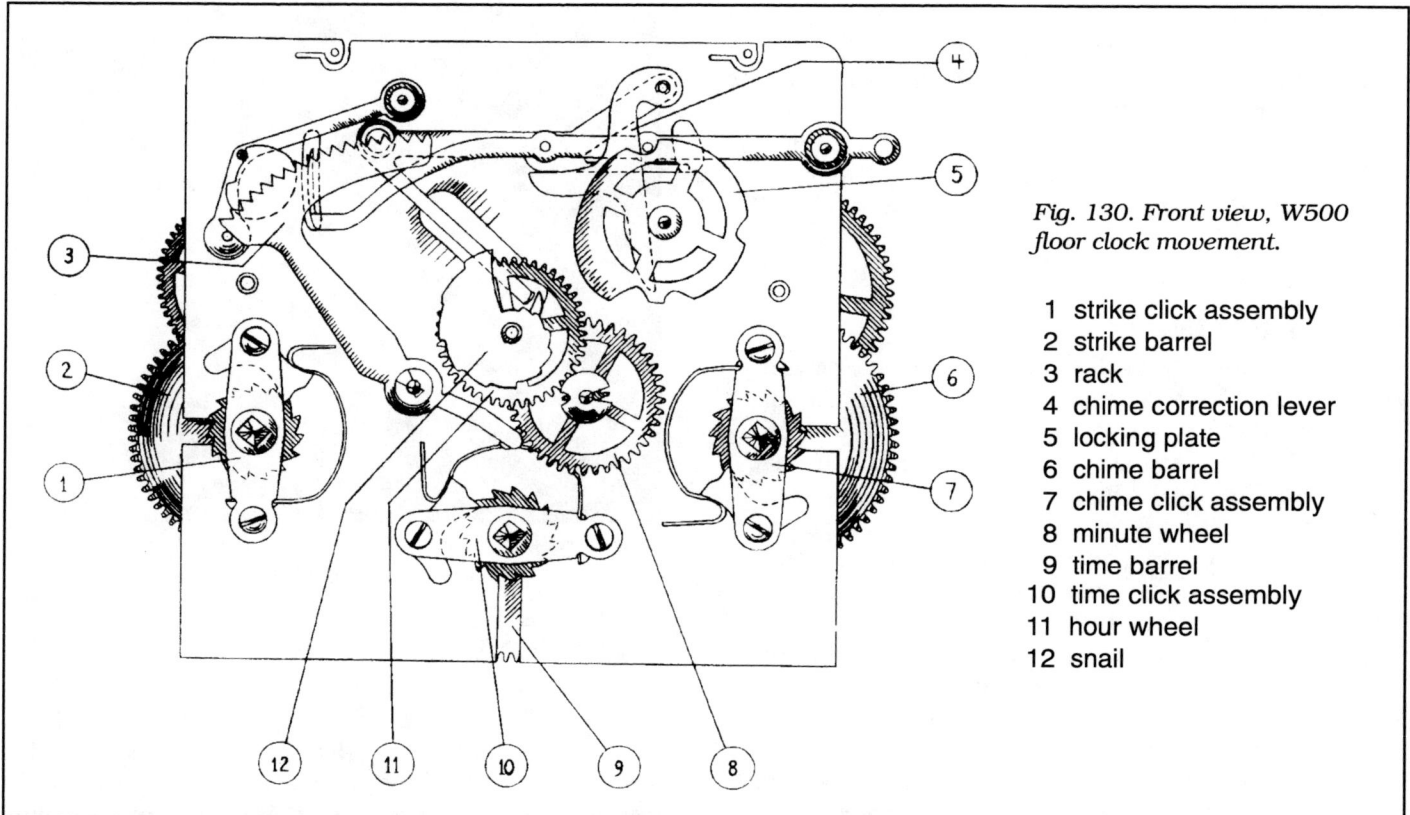

Fig. 130. Front view, W500 floor clock movement.

1 strike click assembly
2 strike barrel
3 rack
4 chime correction lever
5 locking plate
6 chime barrel
7 chime click assembly
8 minute wheel
9 time barrel
10 time click assembly
11 hour wheel
12 snail

and strike parts of the Mauthe W500. Repairing the movement is easier with some notes and diagrams to guide you. The front movement chime and strike parts are the most difficult parts of the job. The locking plate (5) is set-screwed to the front pivot of the third chime arbor, controlling the number of notes chimed at each quarter hour. The chime locking lever (22) stops the gear train at the end of each note sequence by catching the chime lock pin (21) and the locking wheel (20). The strike is rack and snail type. A notched portion of the gathering pallet (16) holds a pin on the rack hook (17) to stop the strike train.

Before you take the movement apart, mark a few parts to help you avoid mixing them up among the three gear trains during reassembly. I suggest marking the second wheels in the time and strike trains, the warning wheels (18 and 23), the two flies, and the mainspring barrels (2, 6, and 9). A "C" for chime, "T" for time, and "S" for strike will be enough. It is

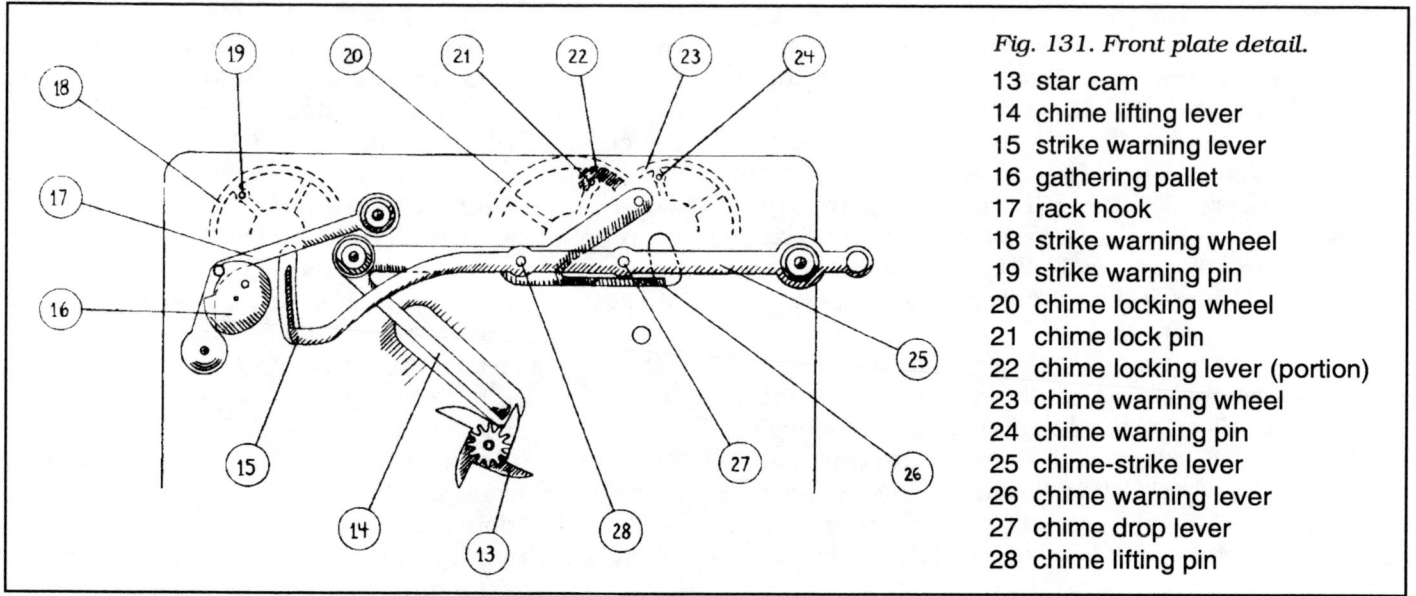

Fig. 131. Front plate detail.

13 star cam
14 chime lifting lever
15 strike warning lever
16 gathering pallet
17 rack hook
18 strike warning wheel
19 strike warning pin
20 chime locking wheel
21 chime lock pin
22 chime locking lever (portion)
23 chime warning wheel
24 chime warning pin
25 chime-strike lever
26 chime warning lever
27 chime drop lever
28 chime lifting pin

not necessary to mark every wheel in the movement. Disassemble the movement for cleaning after the mainsprings are fully let down. The gathering pallet can be driven off after the plates are separated. If no bushing is needed and the pivot is smooth, you may leave it on the arbor. This means the gathering wheel would stay with the front plate.

STRIKE TRAIN

Following cleaning, installation of bushings, and any other repairs, begin reassembly by placing all the wheels into the front movement plate. With the mantel clock version, which has the hammers mounted underneath the movement, the pin barrel and hammer assembly are located between the plates instead of on the back of the movement. Therefore, they must be added at this time. Note that the long pivot on the pin barrel faces the rear on the mantel version, to receive the pin barrel wheel (34, Figure 133). On the floor clock version, the vertical back mounted hammers are added later.

After getting the back plate on and the pivots in their holes, check the strike train first. Add the strike hammer assembly to the back of the floor clock movement. In the mantel clock movement shown in Figure 133, the lift wire (29) is hooked onto the hammer assembly (30), then onto the hammer-lifting lever (31). Finally, the lever is installed and held with a split washer. There is a washer under the hammer-lifting lever which must not be left out.

Next, add the rack hook (17, Figure 131) without the taper pin fastening it in place. Rotate the wheels by hand, with the fly going counterclockwise viewed from the front. The strike hammers should rise and fall once before the train locks. If the hammers are left in the raised position, separate the plates enough to allow changing the mesh of the pin wheel and the pinion on the gathering arbor. Following the adjustment, there must be no load on the hammers when the strike train is locked or when warning occurs. Instead of separating the plates, you can remove the gathering pallet, and try it in each of the other three possible positions. The mantel version allows very little clearance between the hammer-lifting star and hammer tail, making this a touchy adjustment.

The strike warning pin (19) should be at a 10 to 12 o'clock orientation when the strike train is at rest. This will allow for the run of the warning wheel (18). Separate the plates and re-mesh the warning pinion with the wheel below. It would be a good idea to assemble the wheels correctly the first time you put the plates together, but this is difficult to accomplish.

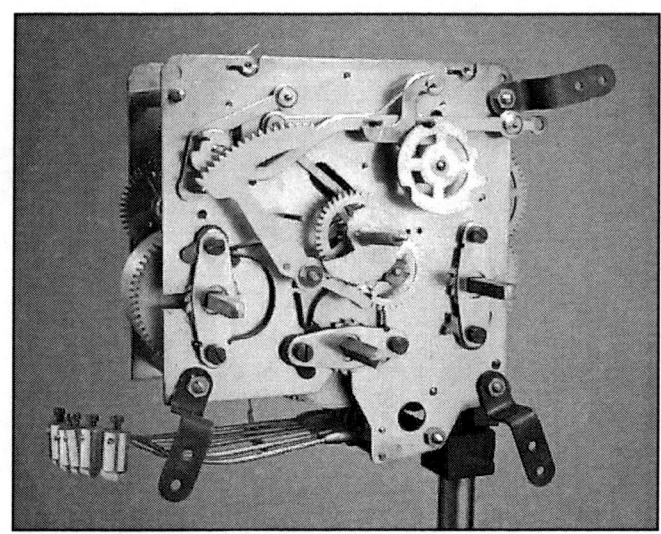

Fig. 132. Front view of the Mauthe W500 IV movement, mantel clock version.

CHIME TRAIN

Check the chime train next, referring to Figures 130 and 131. First, install the chime lifting lever (14) but do not fasten it in place. The chime correction lever (4) can be added, but may be left out until later as it may get in your way. Add the chime-strike lever (25) and the chime locking lever (22). The chime locking lever locks the chime train at the end of each note sequence. The chime-strike lever has several purposes: 1) it has the strike warning lever (15) on the left end; 2) the chime lifting pin (28) is in the center; and 3) the chime drop lever (27) is on the right end. The drop lever rides on the locking plate, and near the end of the hour chime it is lifted high by the locking plate. In this way it serves also as a strike lifting lever, and it brings about the strike warning. At other times, the lever falls into the slots in the locking plate, lowering the chime locking lever enough so that it catches the chime lock pin. To install the chime locking lever, hold it in position between the movement plates. Insert the arbor of the chime-strike lever first through the hole in the front plate, then through the sleeve on the chime locking lever. Rotate the wheels so the fly moves counterclockwise, until the chime locking lever stops the chime lock pin. Now observe the chime warning pin (24). It should be at about a 12 o'clock position for the chime warning setting. Separate the plates slightly and correct the warning wheel (23) if necessary.

Proceed to assemble the front movement parts. Go back to the rack hook and add the taper pin behind the front movement plate. Check the rack hook for freedom of motion. Next, fasten the chime lifting lever by installing the washer on the portion of the arbor extending through the back plate. In-

CHAPTER 19 - MAUTHE W500

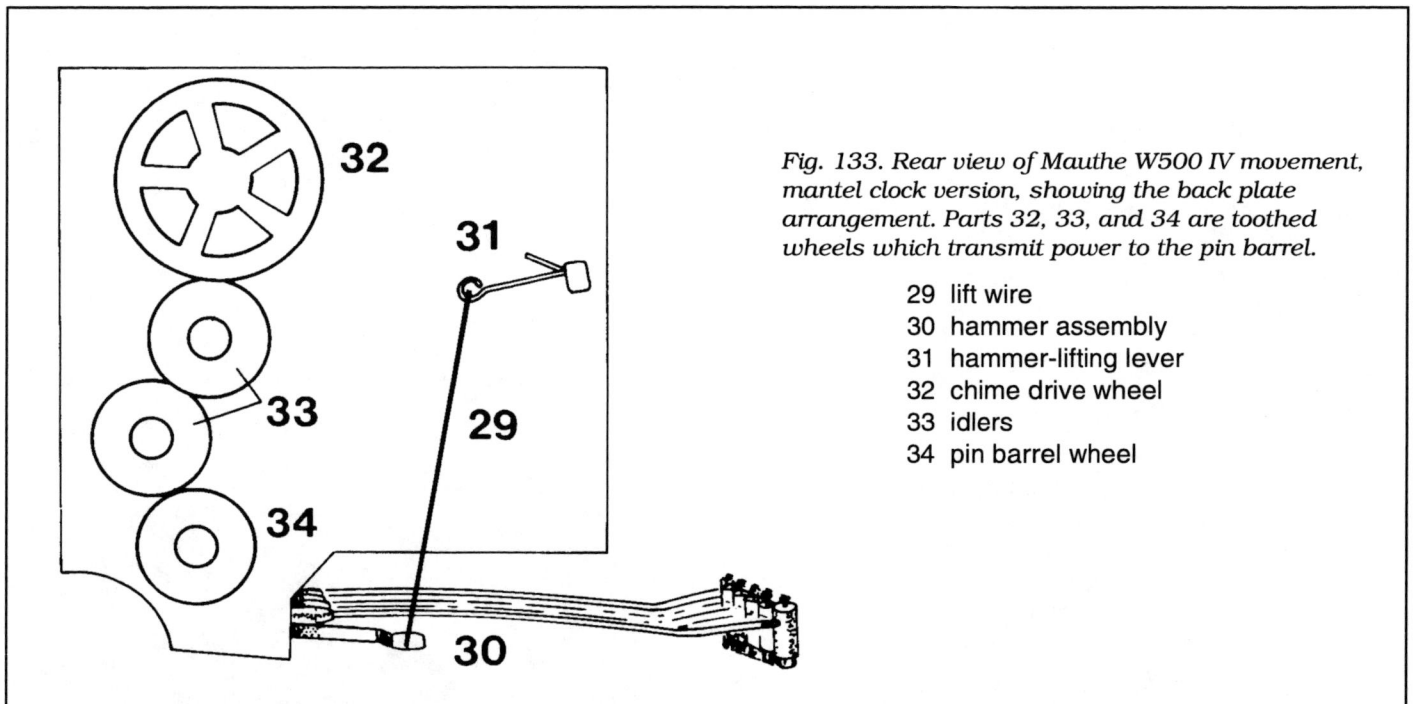

Fig. 133. Rear view of Mauthe W500 IV movement, mantel clock version, showing the back plate arrangement. Parts 32, 33, and 34 are toothed wheels which transmit power to the pin barrel.

29 lift wire
30 hammer assembly
31 hammer-lifting lever
32 chime drive wheel
33 idlers
34 pin barrel wheel

sert a piece of soft wire through the hole in the arbor and bend it into an "s" shape to hold the washer in place. Check to be certain that this arbor also moves freely, without a trace of binding. Add the chime correction lever if it is not already there, fastening it with a small split washer. The locking plate is then installed, with the set screw left loose.

Continue the chime assembly work by turning the wheels in the operating direction until the chime locking lever reaches the pin. Tighten the set screw lightly on the sleeve of the chime locking lever, making sure you leave endshake. More adjustment may be needed in a moment. Turn the locking plate so the chime drop lever falls into one of the four slots. Tighten the locking plate set screw.

If the plates do not need to be separated again for adjustments, it is time to install the mainspring barrels. Refer to Figure 130 for the positions of the three click assemblies. Make sure the pillar nuts are tight before you wind the mainsprings. Put on the minute hand and turn the movement through several quarter hours. If the chime train fails to lock at the end of a chime sequence, or will not unlock to begin chiming, adjust the chime locking lever. Loosen the set screw on the sleeve of the lever, and increase or decrease the depth of locking.

CHIME CORRECTION MECHANISM
The chime correction device is made up of a cam behind the locking plate, and the chime correction lever (4, Figure 130). The lever pivots freely on the end of the chime lifting lever. At the end of the third quarter chime, the chime correction lever comes to rest against the flat side of the cam. This lowers it away from the chime lifting pin (28) on the chime-strike lever. If the next chime point reached by the minute hand is not the hour, the star cam (13) does not raise the lifting lever high enough to start the hour chime. Only the hour position lifts the lever high enough. The chimes can be silent for up to an hour while this correction process takes place. The only adjustment required for the chime correction mechanism would be to adjust the amount of lock on the chime locking lever.

FINISHING THE ASSEMBLY
To complete the assembly of parts onto the front of the movement, install the rack (3), minute wheel (8), and hour wheel (11). Adjust the snail position by meshing the hour wheel to permit correct striking at 12 and 1 o'clock. Install the chime hammers onto the rear of the floor clock movement, with the pin barrel and bracket plate. The hammer assembly should already be in place on the mantel clock version, because it had to be added before the back plate, as described earlier in the chapter.

Adjust the chime note sequence by chiming the clock through the hour and then the first quarter. Loosen the two set screws on the large chime drive wheel on the back of the movement. Turn it counterclockwise with your fingers, as you watch the hammers. When the four chime hammers operate in order from rear to front, tighten the set screws. Make sure that none of the hammers ends a sequence in the raised position.

20

SMITH'S

This chapter covers an English chime movement from a round top mantel clock about 13" wide. The movement is marked:
-SMITH'S-
-ENFIELD-
Made in GT. Britain
by the Enfield Clock Co. Ltd.

Overall movement dimensions are 5-1/4" wide, 5 3/4" high, and 3 1/4" deep (including the center arbor). The clock arrived in the shop with the movement dirty and worn. In addition, the time and chime mainsprings were in need of replacement because of cracked ends. I noticed a previous repair on the time second wheel. One tooth had been removed, and a new one fitted and filed by hand. It was nicely done and no cause for concern. The damage was undoubtedly the result of an earlier mainspring failure. My overall impression of the movement was that it was well constructed and relatively easy to work on.

The Smith's movement is powered by three mainsprings in barrels. The recoil escapement features solid steel pallets rather than the bent strip type. There are five hammers, with the front four for the Westminster chime. The strike mechanism lifts the rear three hammers for the hour count. A chime silencer lever pierces the dial at the numeral "3".

Figure 134 shows a four-hammer chime movement with 5-1/4" square plates, marked "Enfield Clock Co." which has a similar arrangement of chime and strike parts.

STRIKE

The Smith's strike mechanism is a basic rack and snail type. Figure 135 shows the front movement parts. The gathering pallet (4) counts off teeth from the rack (6) during the hour strike. The gathering pallet does not provide the means for locking the strike train at the end of the sequence. Instead, the notch in the gathering pallet allows the rack hook

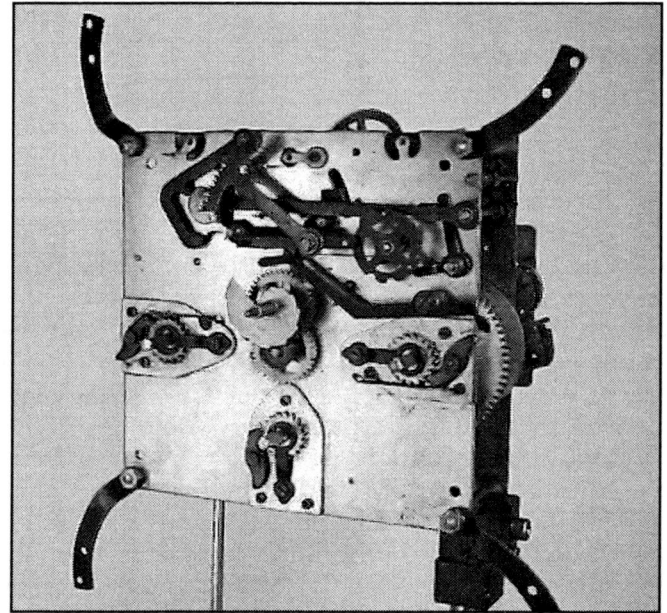

Fig. 134. Front view, Enfield Clock Co. chime movement.

(3) to move downward. The rack hook pulls the strike locking lever (5) with it, because the two are connected. A pin on the rack hook slides in a slot in the lever, joining the two parts. The strike lock pin (not shown) hits the lever to stop the strike train.

Figure 136 shows the means for starting up the strike train at the hour. The long arm on the star cam (12) corresponds to the hour lift. As the lifting piece (19) moves up, the chime lifting pin (17) also moves up. The strike unlocking lever (15) pushes the rack hook, which in turn releases the strike locking lever. Strike warning is "pre-set", so you cannot adjust it. The strike lock pin also serves as the warning pin. When the warning occurs, the pin and wheel simply move to the strike warning lever (14). The pin remains at the strike warning lever until the completion of the chime, when the lever drops out of the way. Striking then proceeds under control of the rack and snail.

CHAPTER 20 - SMITH'S

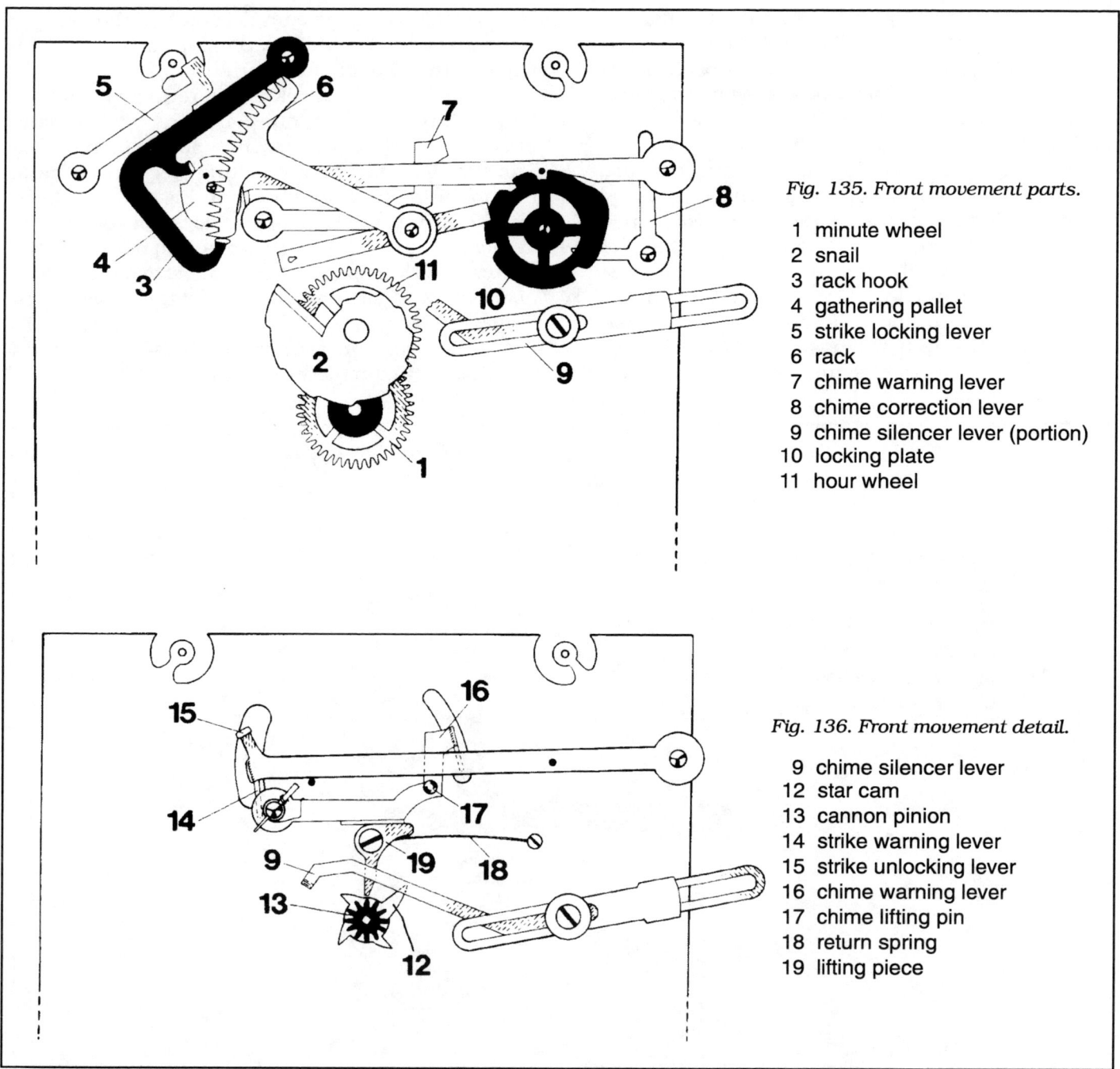

Fig. 135. Front movement parts.

1. minute wheel
2. snail
3. rack hook
4. gathering pallet
5. strike locking lever
6. rack
7. chime warning lever
8. chime correction lever
9. chime silencer lever (portion)
10. locking plate
11. hour wheel

Fig. 136. Front movement detail.

9. chime silencer lever
12. star cam
13. cannon pinion
14. strike warning lever
15. strike unlocking lever
16. chime warning lever
17. chime lifting pin
18. return spring
19. lifting piece

CHIME

The chime mechanism is simpler than it looks. Between the plates, the chime locking system and the chime correction mechanism operate together. It's easy to become confused over them unless you study the mechanisms separately at first. The locking plate is mounted on the front of the movement. It controls the length of each sequence of chime notes, and signals the end of each series. To find the locking action, we have to go behind the front plate. Figures 137 and 138 diagram the parts to be studied. If you locate the next arbor above the one which carries the locking plate, you will find two cams. We are concerned at the moment with the chime locking cam (20) to the rear. The other cam is for chime correction only, and is shown on the next drawing.

Figure 136 shows the lifting mechanism which starts the chime train each quarter hour. The lifting piece (19) raises the long lever across the front of the movement. Moving to Figure 137, the chime locking lever (23) is mounted between the plates on the same arbor as this long lever. As the lifting action takes place, the lever moves up and is lifted out of the cam slot. Warning occurs as the chime warning wheel (21) and chime warning pin (22) move from the rest position down to the chime warning lever (16). The warning action permits the cam to rotate enough to move the slot away from the lever. At the exact quarter, the lever drops back down again, and the gear train runs as long as the locking plate allows. Then the lever drops back down to hold the cam.

The locking plate and the chime locking cam are synchronized so that the chime locking lever drops into the slot in the cam just as the musical sequence ends. The most important thing to note is that the cam and lever actually stop the train. As you analyze the movement, this is one of the first things you would want to know. Comparing the Smith's movement to the Seth Thomas No. 124 in Chapter 3, you will find a cam of similar size and location.

But the Seth Thomas cam does not do the locking: a pin and lever do the job. Similar parts do not always have the same function.

Another part to check is the return spring (18, Figure 136). It comes into play if the minute hand is turned counterclockwise through a chiming point. The spring must bring the lifting piece (19) back to the operating position shown in Figure 136. Without the spring, the lifting piece does not come back, and chiming will not occur from then on. When the center arbor turns clockwise, the return spring has no function to perform. The force of the chime warning lever (16) pushes down on the lifting piece. In Figure 136, you can see a wire spring wound around the chime warning lever to assist.

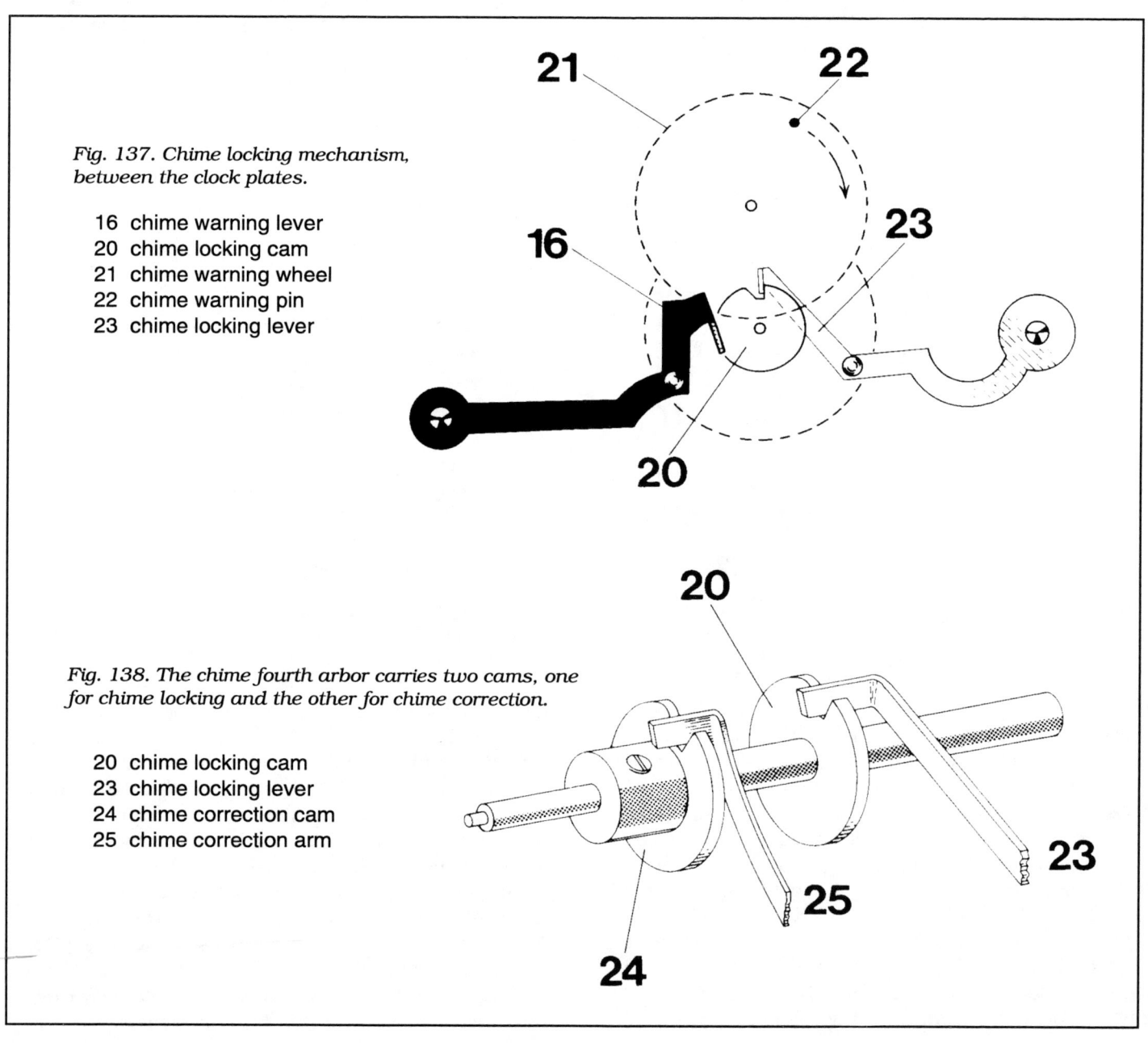

Fig. 137. Chime locking mechanism, between the clock plates.

16 chime warning lever
20 chime locking cam
21 chime warning wheel
22 chime warning pin
23 chime locking lever

Fig. 138. The chime fourth arbor carries two cams, one for chime locking and the other for chime correction.

20 chime locking cam
23 chime locking lever
24 chime correction cam
25 chime correction arm

CHAPTER 20 - SMITH'S

REASSEMBLY AND ADJUSTMENT

As chime movements go, this one is relatively easy to put together. The first step is to fit all the arbors between the plates. Be sure to include the strike hammer arbor at this time, because you cannot add it later without taking the whole movement apart again. This is because the rear of the hammer arbor is elongated to accommodate the strike hammer linkage.

The pillars for the hammer assembly must also be added at the beginning, for the same reason. The threaded portions make them too long to go in later. However, the rear bracket plate for the assembly, along with the pin barrel and hammers, can wait. One thing to note is that there is a brass spacer which must go onto the arbor which carries the hammers. It is easily misplaced during the cleaning process.

Strike Adjustments

My usual procedure, once the plates are together, is to check out the strike first. In this movement, it is straightforward. Figure 135 shows how the rack hook rests in a notch in the gathering pallet. This is the locked position. The strike lock pin will rest against the strike locking lever, to stop the strike train. You must check to see whether the hammer tail has just cleared the hammer-lifting star (not shown). This movement doesn't appear to allow any clearance at the rest position. As soon as the hammer tail drops off one of the points of the star, it falls to touch the next one. So your best adjust-

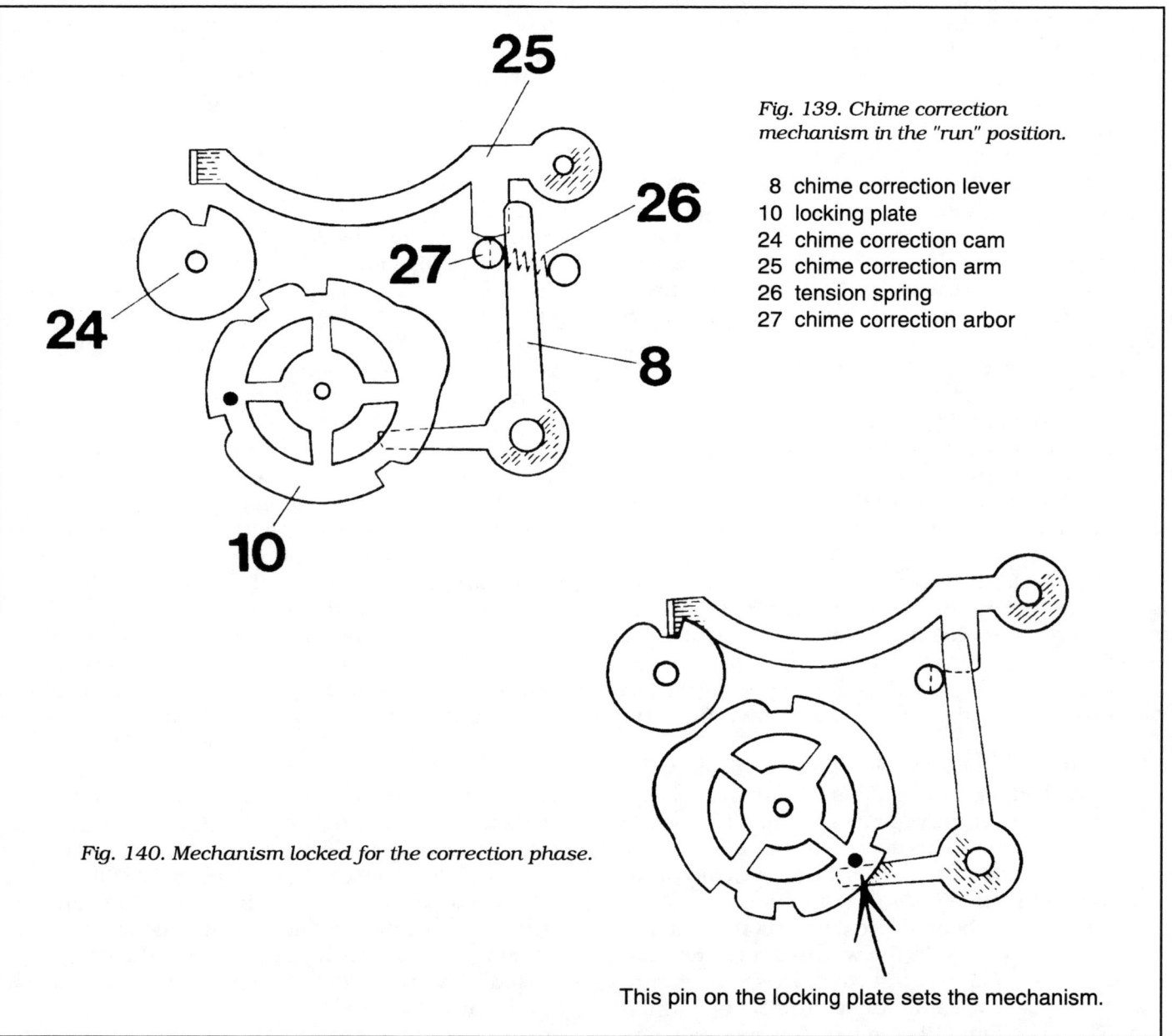

Fig. 139. Chime correction mechanism in the "run" position.

 8 chime correction lever
10 locking plate
24 chime correction cam
25 chime correction arm
26 tension spring
27 chime correction arbor

Fig. 140. Mechanism locked for the correction phase.

This pin on the locking plate sets the mechanism.

ment is to set the train up so it will lock just after this drop occurs. To make an adjustment, make sure all mainspring power remains off. Separate the plates enough to permit you to disengage the pin wheel from the pinion on the gathering arbor. Experiment until you are satisfied.

The strike warning does not have to be set. The strike lock pin just moves the fixed distance from the locking to the warning lever, and that's all.

Chime Adjustments

After adding the locking plate to the front of the movement, rotate the wheels until they lock. At the same time, set the locking plate so that one of the slots is directly under the pin on the long lever across the movement. As shown in Figure 137, the locking action is between the plates on the chime locking cam. Set the warning action as indicated in the drawing, to allow about a half revolution of the warning wheel. This places the chime warning pin at a 1 o'clock orientation when it is at rest. To adjust, you need to separate the plates and ease out the rear pivot of the warning arbor. Reset and check.

The chime drive wheel on the back of the movement carries a set screw. Loosen it and turn the wheel by hand to operate the chime hammers. After you have placed the hammers in sequence with the chimes, tighten the screw. The four chime hammers should move in order from front to rear as they sound the first quarter. I often leave this step for last, after all other adjustments.

Chime Correction Mechanism

Figures 138 through 140 illustrate the main features of the automatic chime correction device. Look first at Figure 138. It shows the chime fourth arbor. As I mentioned earlier, there are two cams on the arbor. The rear one is the chime locking cam (20). The chime locking lever (23) stops the chime train against the slot in this cam. In front of the cam is another one, the chime correction cam (24). It is almost identical in appearance. There is, however, a set screw for adjustment of the chime correction cam. Note the location of the chime correction arm (25).

First of all, the cam must be oriented the same way as the locking cam. That is, the sharp angle of the slot must be at the right as shown. It is possible to have the chime correction cam on backwards, facing the wrong way. If this happens, the automatic self-correction does not function.

The basic idea behind the device is to place the chime correction cam in the way of the chime train before the hour chime. As you can see in Figure 138, it locks in the same way as the chime locking cam. The star cam (12, Figure 136) has only one arm long enough to raise the correction lever out of the cam slot to release the chime train. In operation, the chime train remains silent until the hour chime can play at the hour.

Figure 139 shows the mechanism in the "run" position. The device will not stop the chime train because the chime correction arm (25) is safely above the cam (24). To achieve this position, several parts are required. The chime correction arm rests on the chime correction arbor (27). The arbor runs between the movement plates. In front, it pivots in a wide slot. The tension spring (26) holds the arbor to the right side of the slot.

Figure 140 shows the "correction" position. At the end of the third quarter chime, a pin on the locking plate pushes down on the chime correction lever (8). The arrow in the drawing indicates the pin. This sets the mechanism by causing the lever to push the arbor (27) to the left, against the spring pressure. The chime correction arm drops off the arbor. (The arbor is notched, as indicated by the dotted line through its diameter.) As a result of all this, the chime correction arm drops down as shown in Figure 140. It can stop the gear train to prevent chiming.

Once the chiming does begin on the actual hour, the arm (25) moves up again. Aided by the coil spring, the arbor (27) slides under it to keep it clear of the cam (24). The mechanism is now back to the Figure 139 "run" position until it is set again.

The major adjustment is to the chime correction cam (24). If it is installed backwards, the sloping side of the slot cannot arrest the train. No matter how smoothly the mechanism operates, it will not accomplish anything. Install the cam as shown in Figure 138. It must be set back far enough to clear the chime warning wheel. If necessary, you can also move the chime correction arm (25) further back on its arbor to give yourself more room; there is a set screw to permit the adjustment.

Rotate the chime correction cam (24) to the correct spot. The slots in the two cams (24 and 20) are not supposed to line up exactly with each other. The chime correction cam should be set a few degrees clockwise as shown in Figure 138. This allows the chime locking lever and cam to lock up the chime train routinely each quarter. The chime correction cam and arm provide a secondary lock before the hour only. If the long arm of the star cam is raising the lifting piece, the extra height will be more than enough to overcome this second locking action. If another arm is there, then it is not the hour. Although the chime locking lever and cam are released, the chime correction parts will still hold up the chime.

INDEX

assembling movements, tips on, 4
assembly procedure, general, 4-6
Barwick, 93
Bauerle, 75
bevel gear, in Urgos chime, 102
Borgfeldt, George, 75
brackets, *see movement brackets*
cable drive, 13
Canterbury chime, 2
chime
    automatic changing of, 75, 78-80
    Canterbury, 2
    melodies, 2-3, 75
    rods, 9-10
    selector, 78-79, 89, 102, 108
    silencing, 17, 19, 62, 77, 102
    St. Michael, 2, 3, 98, 108
    Trinity, 75
    Westminster, 2, 3
    Whittington, 2, 3, 98, 108
    Winchester, 2
chime correction mechanism, 21, 23-25, 29, 37, 63, 48-49, 56, 60-61, 68, 71-72, 77, 88-89, 95, 104
clicks, use of double, 27, 53
Coleman, Jesse, reference to, 66
Colonial Mfg. Co., 69, 112
count wheel, 35, 54
Daneker Clock Co., 116
Double Deck Chime, Waterbury, 52-58
dual chime, in Jacques clock, 75, 78
dual rack, in Sessions clock, 47-48
Enfield Clock Co. (Smith's), 120
floating balance, 86
Fried, Henry, references to, 4, 75
Hamburg American Clock Co., 66
hammer
    adjustments, 9-10
    heads, 9
    mechanism, 8, 11
Hermle date codes for movements, 85
Herschede
    winding mechanisms, 12
    producing tubular bell clocks, 97
Kieninger
    cable bracket, 15
    cable drive, 14
    stop-works, 15
locking action, 96
locking plate, 7
mainspring box, 12-13, 16
mainspring sizes
    calculating for barrels, 65
    for Seth Thomas No. 124, 26

    for Seth Thomas No. 113, 27
    for New Haven, 36
    for Sessions, 44
    for Ansonia, 65
    for Urgos 06 serles, 93
Mansion series, Waterbury Clock Co., 55-58
Miller, Howard, 93
movement brackets, 89, 93-94
night shut-off, automatic, 109
pin barrel, 2
pivot surface, 1
quarter rack, in Sessions clock, 47
rack and snail, illustrated, 58, 60
rack let-down, 73
ratchet, see *click*
silencing, see *chime* and *strike*
spring driven movements, 11-13
springs, hammer, 91
St. Michael chime, 2, 3, 98, 108
stop-work mechanism, 15
strike
    adjustments, 5-6
    silencing, 102-103, 111
Thomas, Seth, clocks
    Chime Clock No. 72, 26
    Chime Clock No. 73, 26
    Chime Clock No. 90, 16
    Chime Clock No. 96, 16
Thomas, Seth, movements
    No. 89, part of Sonora chime, 33-35
    No. 113, compared to No. 124, 26
Trinity chime 75
triple chimes, 97, 102-104, 108-109
tubular bells, 4, 69, 75, 97
two-train chime, 44, 66
Uhrwerk, 93
ultrasonic cleaning, 2
Urgos
    03 series, 93
    06 series, 97
warning
    adjustment, 95
    lever, 8
wear
    how to recognize, 7-8
    in hammer tail, 90-91
weight driven clocks, 13-14, 69, 75, 89, 97, 104
weights, In Urgos 9-tube clock, 93
Westminster chime, 2, 5
Whittington chime, 2, 3, 93, 108
Winchester chime, 2
winding mechanism, geared, 11-12, 63, 81

ACKNOWLEDGEMENTS

Many people provided advice, information, and clocks for study during the years I worked on writing the original magazine articles and the 1990 edition of *Chime Clock Repair*. Several of the people listed below also helped with suggestions for the second edition.

American Watchmakers-Clockmakers Institute
Kieninger Clocks, Carl Baessler
Watch & Clock Review
Urgos, Carl Bendorf
Sligh Clocks, Steve DeYoung
Henry B. Fried
Josephine F. Hagans
Arlington Book Company, Tran Duy Ly
Hermle Black Forest Clocks, Helmut Mangold
Dale Milsark
National Association of Watch & Clock Collectors
Stan Palen
Al Stevens
David Kanowski
Burton Rought
Arnold Sussman
Jim Alexander
Richard G. Trepp
Bill Rakes

ISBN 0-9624766-6-8